Man's place in evolution

Second Edition

Natural History Museum Publications

Cambridge University Press

Published by Natural History Museum Publications,
Cromwell Road, London SW7 5BD and the
Press Syndicate of the University of Cambridge,
The Pitt Building, Trumpington Street, Cambridge CB2 1RP
40 West 20th Street, New York, NY10011-4211, USA
10 Stamford Road, Oakleigh, Melbourne 3166, Australia

First edition 1980
Second edition 1991

Printed in Great Britain by Jolly and Barber Ltd,
Rugby, Warwickshire

British Library Cataloguing in Publication Data
Man's place in evolution
1. Man. Evolution. For children
I. British Museum (Natural History) Department of Public Services
573.2

Library of Congress Cataloging-in-Publication Data available

ISBN 0 521 40864 4 paperback

Contents

Preface 5

Chapter 1
Man's living relatives 7

Chapter 2
Working out relationships 13

Chapter3
Man's closest living relatives 21

Chapter 4
Man's fossil relatives 27
the first apes, the sivapithecines
and the australopithecines

Chapter 5
Man makes tools 49
the habilines

Chapter 6
Man uses fire 61
the *Homo erectus* people

Chapter 7
Man has ceremonies 71
the neandertals

Chapter 8
Man uses symbols 83
modern humans

Chapter 9
Man has evolved 93

Further reading 97

**Names of fossils
featured in this book** 98

Index 101

Acknowledgements 102

Preface

Man – *Homo sapiens* – is only one of many thousands of animal species alive today. How are we human beings related to other living animals? And how are we related to the various 'fossil men' whose remains have been found? This book will help you to decide for yourself.

It begins by explaining how animals can be grouped together on the basis of the characteristics that they share, and shows that all human beings are vertebrates … mammals … primates … and apes. It then describes a simple method for working out the relationship between different animals, and applies this method to the apes in an attempt to discover which are our closest living relatives.

The book then turns to 'fossil man', but it makes no attempt to reconstruct the history of human evolution. Instead, it looks more closely at man's unique characteristics, and then examines the fossil remains for evidence of these characteristics. In this way a picture is built up of how modern human beings might be related to the ancient sivapithecines and australopithecines, and to extinct human beings such as the habilines and the neandertals. There are many diagrams, colourful reconstructions and photographs – including photographs of classic fossil specimens from all over the world.

This is a companion book to the exhibition **Man's place in evolution**, which opened at the Natural History Museum in May 1980, and was updated in 1986. Like the exhibition, it was planned with the guidance of Museum experts. A great many people, both within the Museum and outside, have helped in the preparation of **Man's place in evolution**, and I should like to take this opportunity of thanking everyone concerned.

Neil Chalmers

Dr. NEIL CHALMERS Director
The Natural History Museum
1991

Chapter 1

Man's living relatives

Most scientists believe that all living things are descended from a single common ancestor. If this is so, all living things – including human beings – must be related to each other.

We can work out how closely human beings are related to other animals by looking at the characteristics we share with them.

9

Human beings are vertebrates.
Like all vertebrates we have...

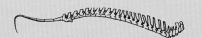

● a backbone

Human beings are mammals.
Like all mammals we have...

● fur or hair

● milk-producing glands

● three separate bones in the middle ear

Human beings are primates.
Like all primates we have...

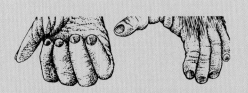

● fingernails and toenails

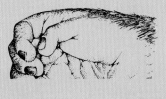

● an opposable thumb...or big toe

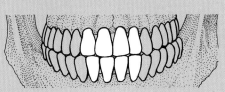

● four incisors in the upper and lower jaw

Human beings are apes.
Like all apes we have…

- no tail

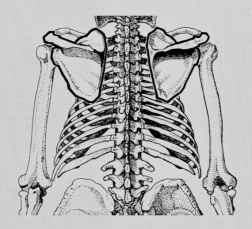

- shoulder blades at the back, not at the sides

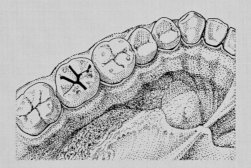

- a Y-shaped pattern on the surface of the molars (chewing teeth)

There are other, more detailed characteristics that are unique to apes. These characteristics are used to find out how closely the living apes are related to one another.

In the next chapter you can discover how biologists work out these relationships.

We cannot go back in time and see how animals evolved, so we can never be sure how the apes – or any other animal species – are related to each other. But we can suggest possible relationships, and then test them to find out which one is the most likely.

How do we begin?

When we are trying to work out the relationships between different species, it is easier if we begin by making two assumptions.

First, we assume that new species arise when one species splits into two, like this

This assumption allows us to test the relationships we suggest, because it means that every species must have a 'closest relative'.

Second, we assume that none of the species we are considering is the immediate ancestor of any of the others. This is a fairly safe assumption, because so many animals have lived and died during the history of life on Earth that the chances of finding – and recognizing – any particular ancestral species are very, very small.

So, if we are considering two species

we assume that they are related like this

(The broken circle represents their ancestor.)

and *not* like this

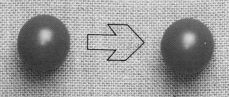

And if we are considering three species

we assume that they are related like this

(The broken circles represent the ancestors.)

In other words, we assume that two of the species share a common ancestor that is not shared by the third species.

This relationship can be represented by a simple branching diagram called a **cladogram.**

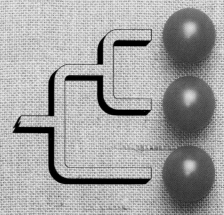

Possible relationships

Let's call our three species A, B and C. They could be related to each other in any one of the following three ways.

A and B could be more closely related to each other than either of them is to C.

This relationship can be shown like this

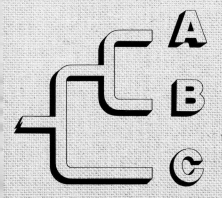

or like this

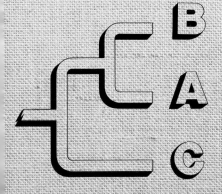

2 A and C could be more closely related to each other than either of them is to B.

This can be shown like this

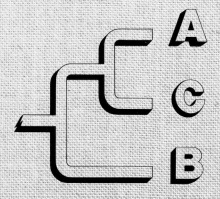

or like this

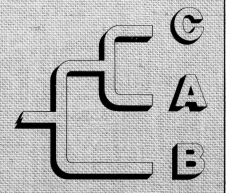

3 B and C could be more closely related to each other than either of them is to A.

This can be shown like this

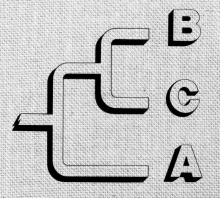

or like this

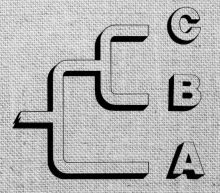

We should test each of these possible relationships, **1**, **2** and **3**, to see which one is most likely.

How do we test relationships?

We test relationships between species by looking for characteristics that are similar in different species because they have been inherited from a common ancestor. These are called **homologies**.

The more homologies that two species share, the more closely they are related. And two species that are 'closest relatives' share homologies that are not shared by any other species. In other words, they share unique homologies.

Looking for homologies

We can look for homologies in a wide variety of characteristics, ranging from whole organs to molecules.

1 Bones and teeth

Characteristics of bones and teeth can provide good examples of homologies. One of the best examples is the backbone that all vertebrate animals share.

Homologies in bones and teeth are very important when we are considering fossil animals, because bones and teeth are often the only parts of the animal that are preserved by fossilization.

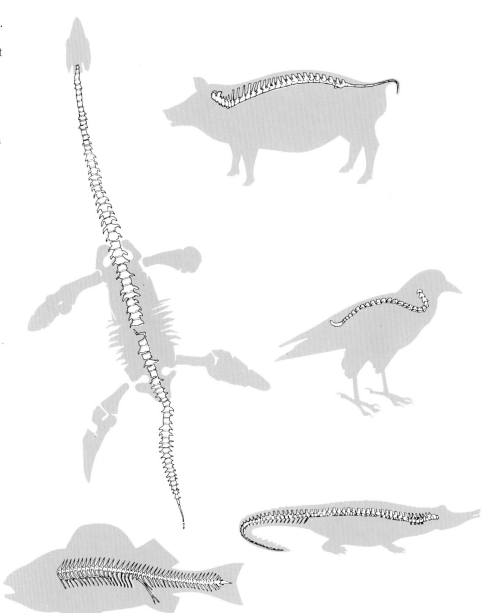

Another homology, shared by some of the apes, is the frontal sinus – a cavity in the skull, just above the eye. Human beings, chimpanzees and gorillas all share this homology, but gibbons, orang-utans and most other primates do not have a true frontal sinus.

But we must be careful. Not all similarities of bones and teeth are homologies. Some similarities may have evolved independently. They may be adaptations to a similar way of life – the wings of bats, birds and pterodactyls, for example. Such similarities cannot be used to test relationships.

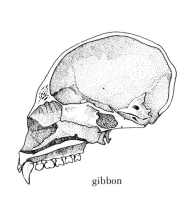

gibbon

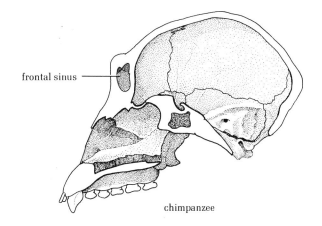

frontal sinus

chimpanzee

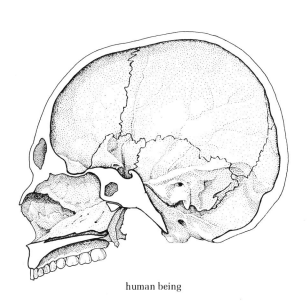

human being

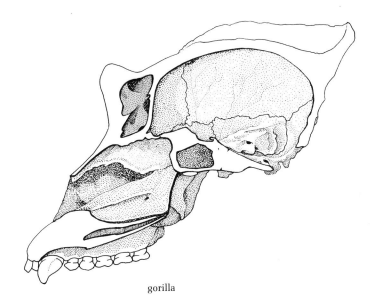

gorilla

2 Soft parts of the body

Homologies can also be found in the soft parts of the body. The milk-producing glands of mammals are a good example.

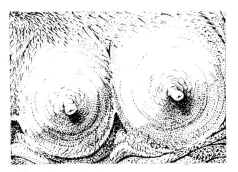

milk-producing glands

Another example is the sitting pads of tough skin ('ischial callosities') that most monkeys have on their rumps. Gibbons also share this homology, but the other apes do not.

But of course not all similarities between the soft parts of different animals are homologies. The parts may be similar because they have a similar function and *not* because they are inherited from a common ancestor. An example is the webs between the toes of ducks, frogs and seals.

3 Chromosomes

At a microscopic level, the chromosomes contained in every animal cell can provide examples of homologies. Chromosomes vary in number and structure between different types of animals. Human beings have 46 chromosomes (23 pairs) in each body cell. Chimpanzees and gorillas have 48 (24 pairs).

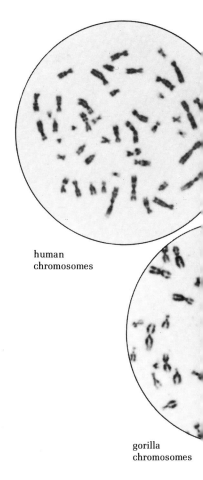

human chromosomes

gorilla chromosomes

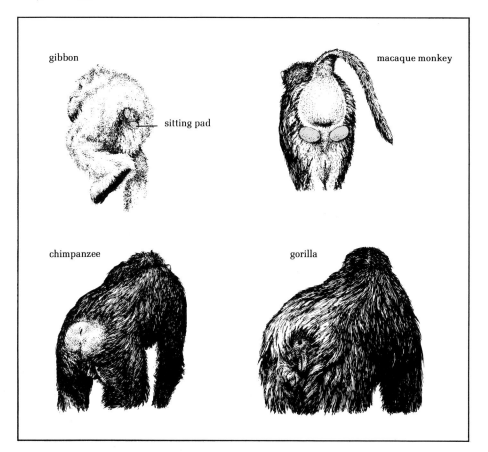

gibbon

sitting pad

macaque monkey

chimpanzee

gorilla

Similarities between the chromosomes of different types of animals probably indicate a common ancestor. It is very unlikely that the similarities evolved independently in each species, and the chromosomes look like each other just by chance.

4 Molecules

We can also find examples of homologies in molecules. DNA, the genetic material contained in chromosomes, is a long chain molecule. Using various techniques the DNA of any two species can be compared.

Other examples can be found in protein molecules, which are composed of amino acid units. The alpha chain, part of the haemoglobin molecule found in red blood cells, is made up of 141 amino acids.

Human beings and chimpanzees have identical alpha chains. But, in other primates, some of the amino acids are different, or are in a different order along the chain.

When two or more species share the ability to make a complex molecule like this, it is almost certain that they have inherited this ability from their common ancestor. Making molecules is such a complicated process that it is very unlikely that the same complex molecule would evolve independently in different species.

chimpanzee
chromosomes

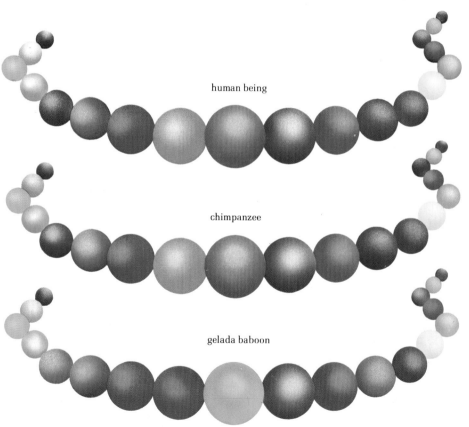

human being

chimpanzee

gelada baboon

These diagrams represent the sequence of amino acids in just one part of the alpha chain.

We can use these different kinds of homologies to test our ideas about the relationships between different species. The more homologies we find, the more confident we are that our ideas are correct.

Chapter 3

Man's closest living relatives

Which of the other apes are our closest relatives? . . . the gibbons, the orang-utans, the gorillas or the chimpanzees?

To answer this question, we must find out if any of the other apes share unique homologies with human beings. If they do, then they must be our closest relatives (see page 16).

Our close relatives

Evidence suggests that gibbons and orang-utans are not as closely related to us as chimpanzees and gorillas are.

For example, human beings, chimpanzees and gorillas all have a frontal sinus (see page 17). Gibbons, orang-utans and other primates do not have a true frontal sinus. This is evidence that chimpanzees and gorillas are more closely related to us than gibbons and orang-utans are.

And evidence from tests on certain blood proteins (fibrino-peptides) suggests that orang-utans are more closely related to us than gibbons are.

We can represent these relationships by a cladogram like this . . .

human beings, gorillas chimpanzees

orang-utans

gibbons

white-handed gibbon
Hylobates lar

orang-utan
Pongo pygmaeus

Our closest living relatives

Chimpanzees and gorillas are more closely related to us than any other living animals are. But which of them are our *closest* living relatives?

chimpanzee
Pan troglodytes

gorilla
Gorilla gorilla

Is it the gorillas?

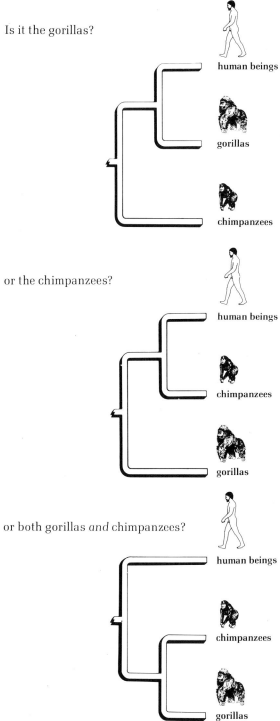

human beings

gorillas

chimpanzees

or the chimpanzees?

human beings

chimpanzees

gorillas

or both gorillas *and* chimpanzees?

human beings

chimpanzees

gorillas

To try to answer this question, we must look carefully at human beings, gorillas and chimpanzees for evidence of shared characteristics.

The traditional view is that chimpanzees and gorillas are more closely related to each other than either of them is to human beings (cladogram **3**). This view is based on similarities between their bones, teeth and soft parts of the body.

But, evidence from studies on chromosomes and molecules has challenged the traditional view.

In assessing the evidence, we must be cautious. Shared characteristics are not necessarily homologies – sometimes they are merely adaptations to a similar way of life.

And to make sure that we are considering only unique homologies, we must check to see whether or not they are shared by other primates too. If a characteristic is shared by other primates (marked by an asterisk * in the table) it is obviously not unique. So it cannot be used to sort out the relationships between human beings, chimpanzees and gorillas.

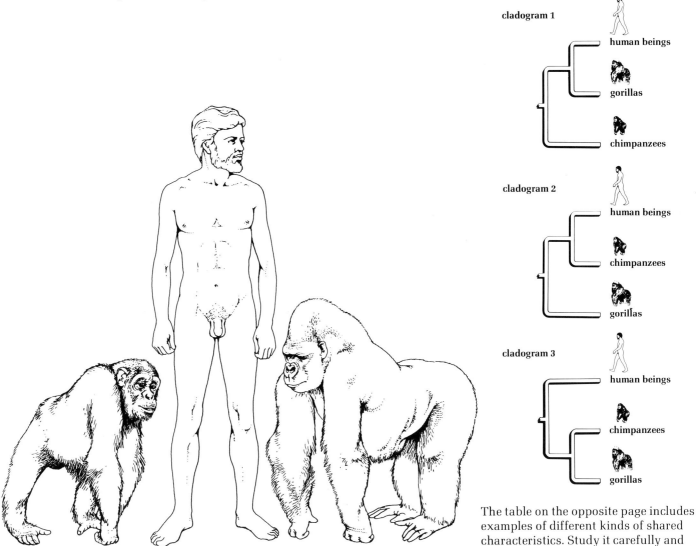

chimpanzee   human being   gorilla

cladogram 1

human beings

gorillas

chimpanzees

cladogram 2

human beings

chimpanzees

gorillas

cladogram 3

human beings

chimpanzees

gorillas

The table on the opposite page includes examples of different kinds of shared characteristics. Study it carefully and see if you can decide for yourself which are our closest relatives.

Shared characteristics	Gorillas	Human beings	Chimpanzees	Other primates	According to this evidence, which cladogram is correct?
Bones and teeth					
limb length	legs shorter than arms	arms shorter than legs	legs shorter than arms	arms and legs equal	3
canine teeth *	large	small	large	large	could be **1, 2** or **3**
thumbs	short	long	short	long	3
Soft parts of the body					
head hair *	short	long	short	short	could be **1, 2** or **3**
calf muscles *	small	large	small	small	could be **1, 2** or **3**
buttocks *	thin	fat	thin	thin	could be **1, 2** or **3**
Chromosomes					
total number	48	46	48	42 or more	3
structure of chromosomes 5 and 12	similar to chimpanzees	similar to other primates	similar to gorillas		3
structure of chromosomes Y and 13	similar to human beings	similar to gorillas	like other primates		1
Molecules					
DNA, compared with that of humans	very similar		very similar	similar	could be **2** or **3**
alpha-haemoglobin chain, compared with that of humans	one amino different acid		identical	several differences	2
sequence of amino acids in myoglobin (muscle protein), compared with that of humans	similar to chimpanzees		similar to gorillas	generally like human beings	3

On the basis of the evidence in the table, which do you think are our closest living relatives? Chimpanzees, gorillas, or both?

Obviously it is very difficult to decide and, whatever choice you have made could well be the right one. On balance the evidence appears to suggest that cladogram 1 is the least likely relationship – but we can get little further.

Even if the table had included all the evidence available, it would not have been possible to decide because much of the evidence is conflicting and the picture is not yet complete.

At present, we are not sure which are our closest relatives. But, for the sake of argument, throughout the rest of this book we will assume that our closest living relatives are both chimpanzees and gorillas.

The next question is
Do we have any closer relatives amongst the fossils?

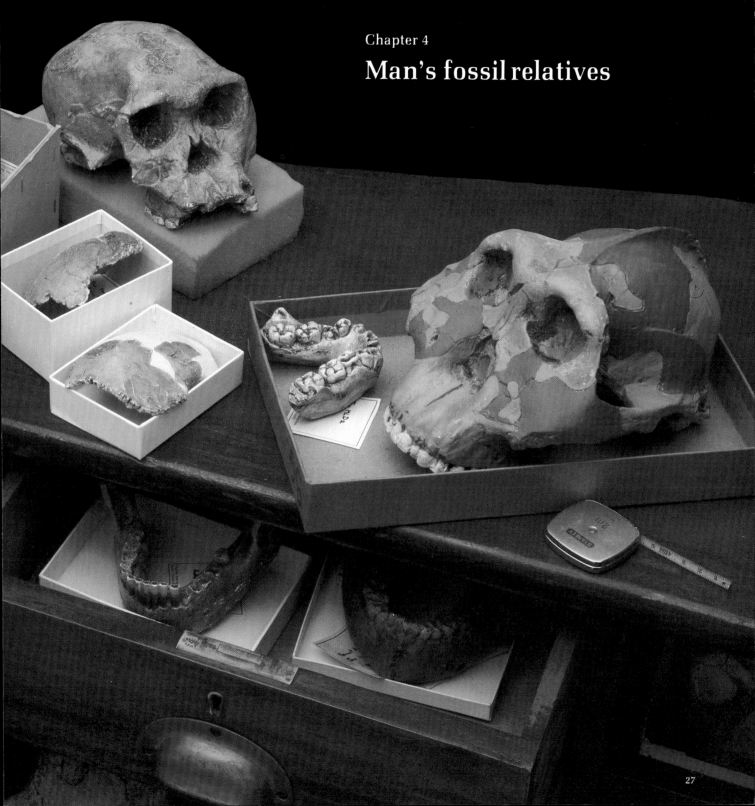

Chapter 4

Man's fossil relatives

The fossilized remains of many different kinds of extinct apes have been found in Africa, Europe and Asia.

How are these extinct apes related to living apes?

The first apes
Ape-like primates were evolving in Africa at least 20 million years ago.

This skull of ape-like *Proconsul* was unearthed in 1948. More recently, several incomplete skeletons of *Proconsul* have been found on Rusinga Island in Kenya.

It is thought that animals similar to this are our oldest ape ancestors.

Skull of *Proconsul africanus*
18 million years old
Discovered in Rusinga Island, Kenya

Reconstruction of
Proconsul africanus

How is *Proconsul* related to living apes?

Proconsul shares the three main characteristics of living apes:

- no tail

- shoulder blades at the back

- a Y-shaped pattern on its molar teeth

However, *Proconsul* also has characteristics that are not found in living apes:

- a shelf or 'cingulum' on its molar teeth

- only one thickened ridge on the inside its lower jaw – living apes have two

These characteristics suggest that *Proconsul* had already diverged from the line which gave rise to living apes.

We can show the likely relationship of *Proconsul* to living apes by a cladogram:

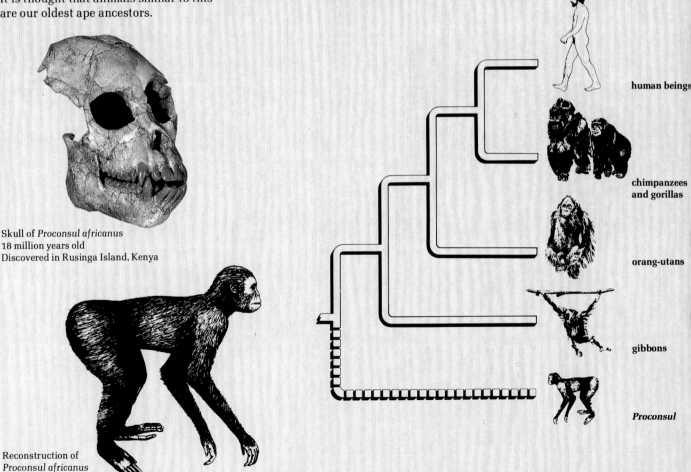

human beings

chimpanzees and gorillas

orang-utans

gibbons

Proconsul

28

The first great apes?

African gorillas and chimpanzees, and orang-utans from east Asia, are known as the 'great apes' – a name separating them from gibbons.

The fossils shown here are important as they may represent the first great apes...

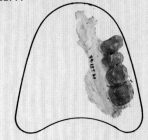

Upper jaw of *Heliopithecus leakeyi*
18 – 17 million years old
Discovered in Ad Dabtiyah, Saudi Arabia

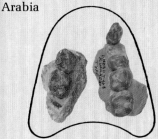

Fragments of upper jaw of *Kenyapithecus wickeri*
14.5 – 12 million years old
Discovered in Fort Ternan, Kenya

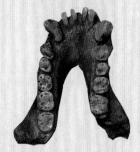

Part of lower jaw of *Dryopithecus fontani*
12 – 11 million years old
Discovered in St Gaudens, France

We are not sure how these fossil apes are related to each other. But we think that as a group they are more closely related to living great apes and human beings than to gibbons. Similarities with great apes and humans include...

● the shape and size of premolar teeth

● thickening of enamel on molar teeth

Again, the relationships can be shown by a cladogram:

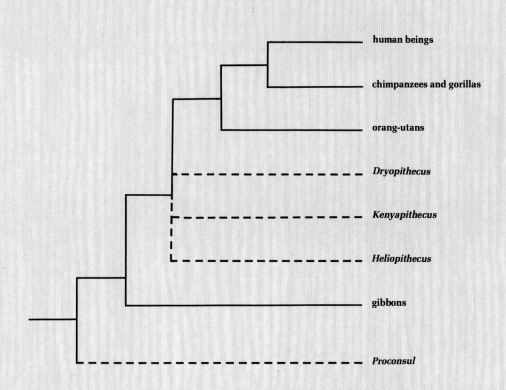

The sivapithecines

The sivapithecines lived between 14 million and 8 million years ago. The name **sivapithecine** comes from *Siva*, an Indian god, and *pithekos*, which is Greek for ape.

These apes had thickened enamel on their teeth, suggesting that they belong to the great ape group.

Their remains have been found in different parts of the world...

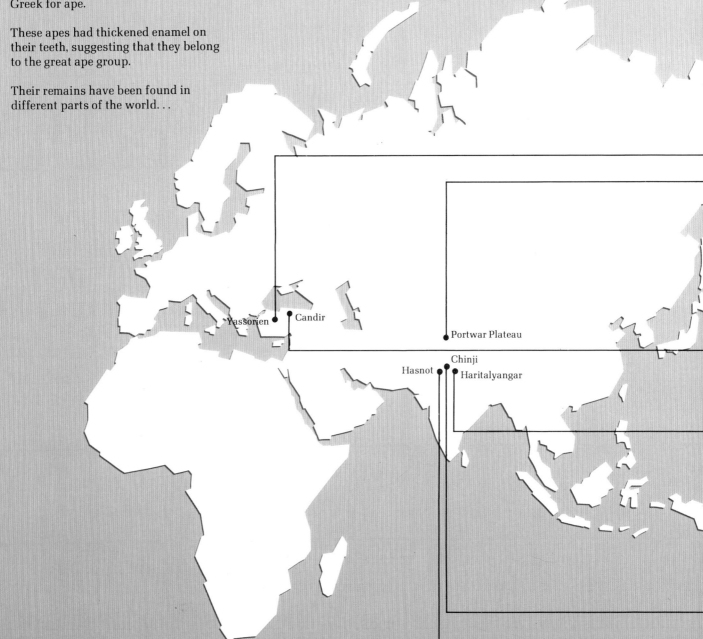

Yassören   Candir

Portwar Plateau

Chinji

Hasnot   Haritalyangar

This group of fossil apes includes several
different species – some much smaller than others.

These are the large sivapithecine apes.

 Lower face of *Sivapithecus meteai*
About 9 million years old
Discovered in Yassorien, Turkey

 Face and lower jaw of *Sivapithecus
sivalensis*
About 8 million years old
Discovered in Potwar Plateau, Pakistan

These smaller apes were once called ramapithecines.
Most scientists agree that they are so similar to the large
sivapithecines that they can now be combined as one group.

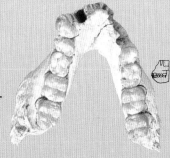

Lower jaw of *Sivapithecus alpani*
About 14 – 12.5 million years old
Discovered in Çandir, Turkey

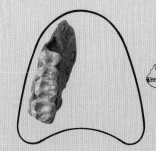

Upper jaw of *Sivapithecus punjabicus*
Originally called *Ramapithecus
brevirostris*
About 9 million years old
Discovered in the Nagri formation,
Haritalyangar, India

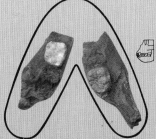

Lower jaw of *Sivapithecus punjabicus*
Originally called *Dryopithecus
punjabicus*
About 8 million years old
Discovered in Chinji, Pakistan

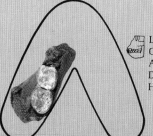

Lower jaw of *Sivapithecus punjabicus*
Originally called *Bramapithecus thorpei*
About 9 – 8 million years old
Discovered in the Nagri formation,
Hasnot, Pakistan

How are the sivapithecines related to living great apes and human beings?

Here is the reconstructed face of *Sivapithecus sivalensis* – the most complete sivapithecine ever found.

Compare it with the skulls of an orang-utan, a chimpanzee and a human being.

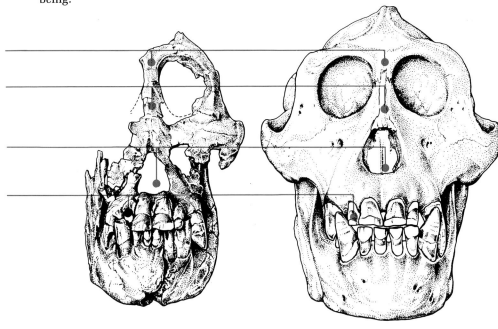

no frontal sinus inside

eye sockets tall, and close together

narrow nose with smooth floor

2nd incisor much smaller than 1st

sivapithecine

orang-utan

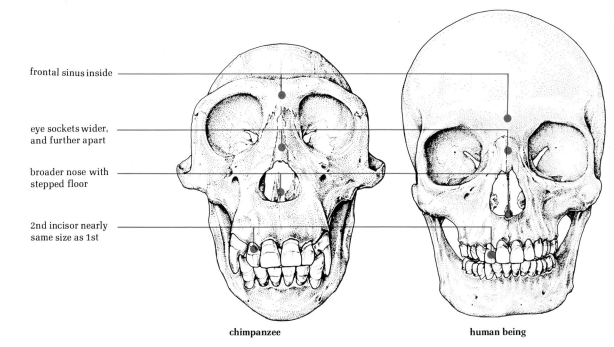

frontal sinus inside

eye sockets wider, and further apart

broader nose with stepped floor

2nd incisor nearly same size as 1st

chimpanzee

human being

As you can see, sivapithecines share many similarities with orang-utans – similarities that no other living apes share.

When we looked at living apes, we separated orang-utans from gorillas, chimpanzees and human beings. The evidence from comparing skulls suggests that sivapithecines are most closely related to orang-utans.

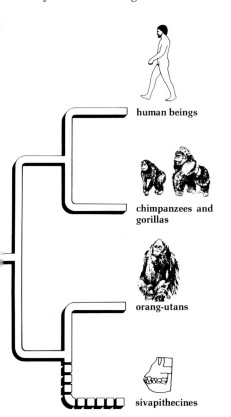

human beings

chimpanzees and gorillas

orang-utans

sivapithecines

The story so far
We can summarize all these relationships in one diagram. . .

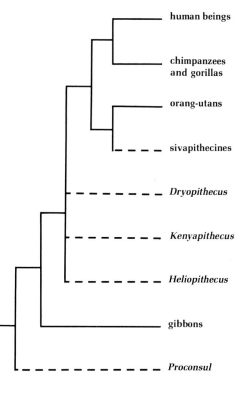

human beings

chimpanzees and gorillas

orang-utans

sivapithecines

Dryopithecus

Kenyapithecus

Heliopithecus

gibbons

Proconsul

So far, fossil evidence has helped us to understand our relationship to other living apes.

But are there any fossil apes more closely related to us than gorillas and chimpanzees are?

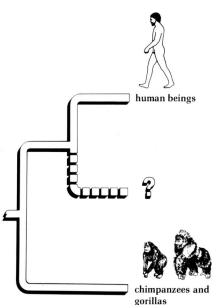

human beings

?

chimpanzees and gorillas

To answer this, we must look for evidence of a fossil ape that shares characteristics with human beings – characteristics *not* shared by gorillas and chimpanzees.

33

The australopithecines

The australopithecines lived between about 5 million and 1.2 million years ago. Their fossil remains have been found in many parts of Africa, especially along the Rift Valley.

The name australopithecine comes from *australis*, which is Latin for southern, and *pithekos*, which is Greek for ape.

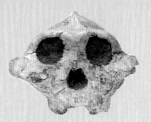

Part of the skull of an australopithecine from Ileret, East Turkana, Kenya

• Hadar

Omo •
• Ileret
Lothagam • • Koobi Fora
• Kanapoi
• Chesowanja
• Peninj
• Olduvai Gorge
• Laetoli

Part of the skull of an australopithecine from Olduvai Gorge, Tanzania

Part of the skull of an australopithecine from Swartkrans, South Africa

• Makapansgat
• Kromdraai
• Swartkrans and Sterkfontein

• Taung

The first australopithecine found. Part of the skull and the brain cast of a young australopithecine found in 1924 at Taung, South Africa

• australopithecine remains found here

How are the australopithecines related to living great apes and humans?

Look closely at the australopithecine skull below and compare it with the human skull and the two skulls of living apes.

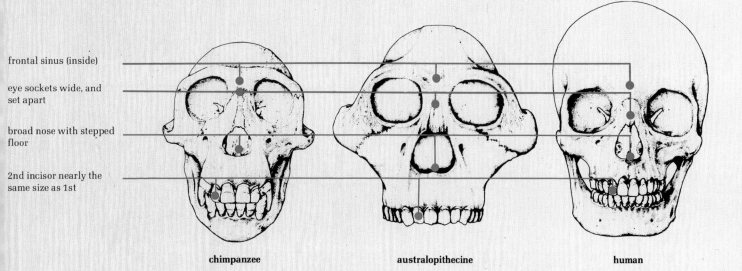

frontal sinus (inside)

eye sockets wide, and set apart

broad nose with stepped floor

2nd incisor nearly the same size as 1st

chimpanzee   **australopithecine**   **human**

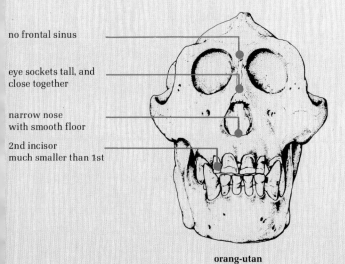

no frontal sinus

eye sockets tall, and close together

narrow nose with smooth floor

2nd incisor much smaller than 1st

orang-utan

You can see that the australopithecine skull has a number of features in common with the human and chimpanzee skulls. This suggests that the australopithecines are more closely related to the human being/ gorilla/ chimpanzee group than to the orang-utan/ sivapithecine group.

But which are *more* closely related to human beings?

chimpanzees and gorillas?

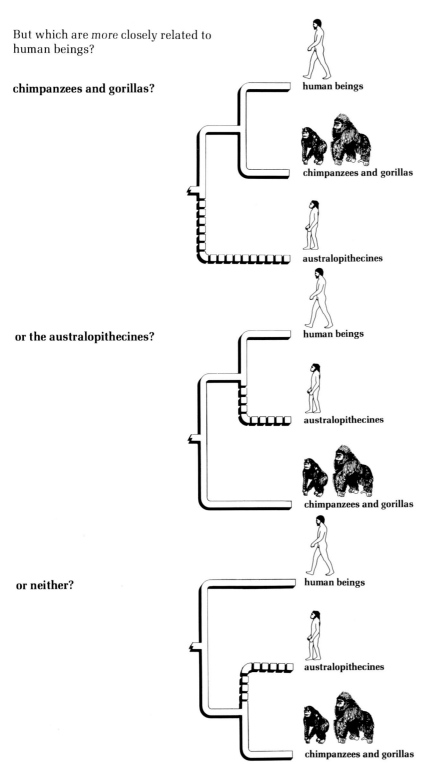

human beings

chimpanzees and gorillas

australopithecines

or the australopithecines?

human beings

australopithecines

chimpanzees and gorillas

or neither?

human beings

australopithecines

chimpanzees and gorillas

To answer this question, we must find out if the australopithecines share any of our special characteristics that gorillas and chimpanzees do not share.

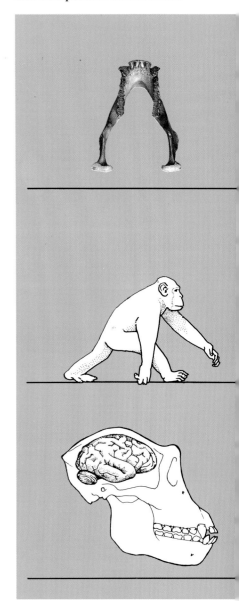

Unlike chimpanzees and gorillas, human beings

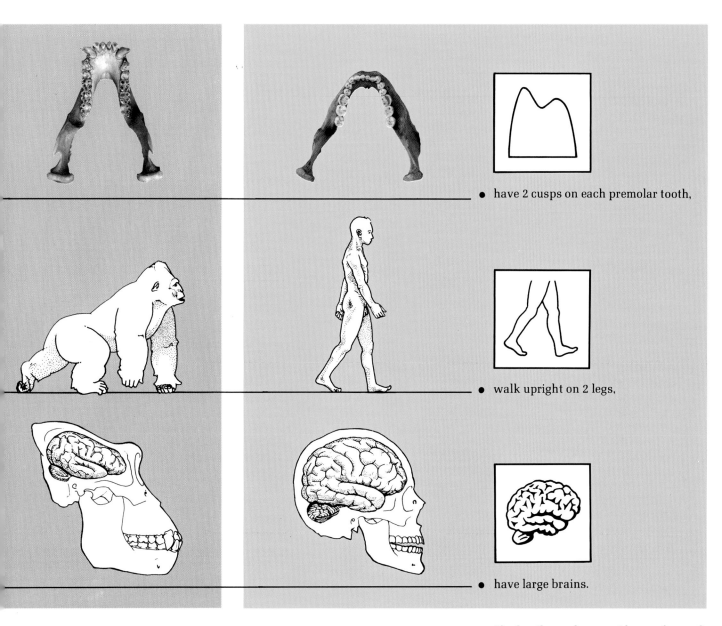

- have 2 cusps on each premolar tooth,

- walk upright on 2 legs,

- have large brains.

If a fossil ape shows evidence of any of these characteristics, we can be fairly certain that it is more closely related to us than chimpanzees and gorillas are.

Look carefully at these jaws – especially at the cusp pattern on the surface of the premolar teeth.

Which ape has teeth most like humans'?

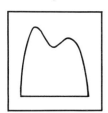

Both the australopithecines and humans have two cusps on each premolar tooth. But chimpanzees (and gorillas) have only one cusp on their lower first premolars.

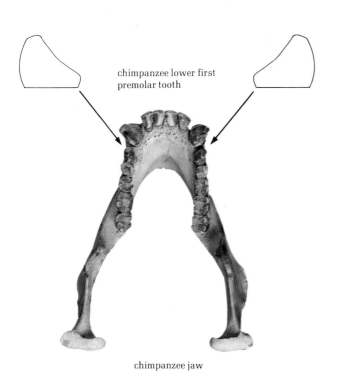

chimpanzee lower first premolar tooth

chimpanzee jaw

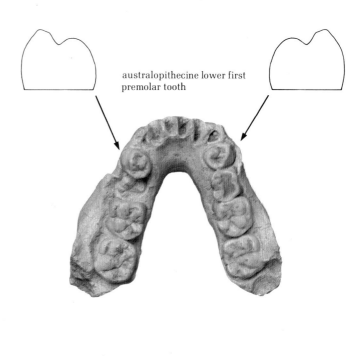

australopithecine lower first premolar tooth

jaw of a young australopithecine from Makapansgat, South Africa

This supports the idea that the australopithecines are more closely related to us than the chimpanzees and gorillas are...

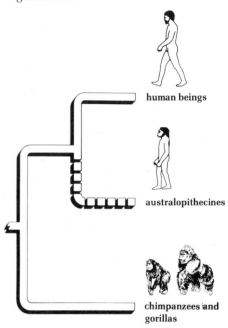

human beings

australopithecines

chimpanzees and gorillas

Is there any other evidence to suggest that the australopithecines are closely related to human beings? Fortunately, fragments of most parts of the australopithecine skeleton have been found. So we can look for evidence of another of our important characteristics – walking upright on two legs.

human lower first premolar tooth

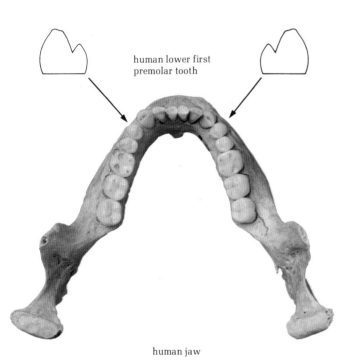

human jaw

**How did the
australopithecines walk?**

Did they walk on all fours
like monkeys?

Or did they knuckle-walk
like gorillas and chimpanzees?

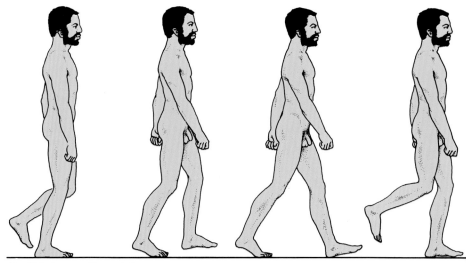

Or did they walk upright
like human beings?

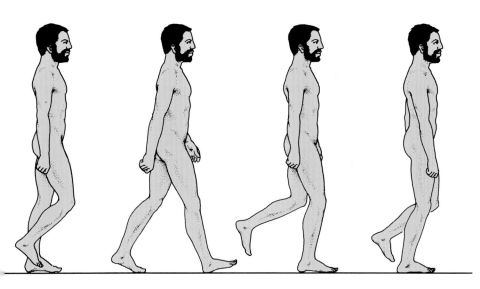

The earliest apes probably walked on all fours. But human beings and gorillas have each evolved their own special way of walking. You can see evidence for their different ways of walking if you compare their skeletons. . .

The skull

human being

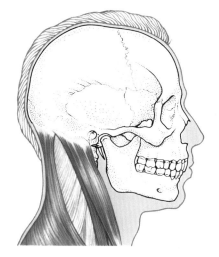

gorilla

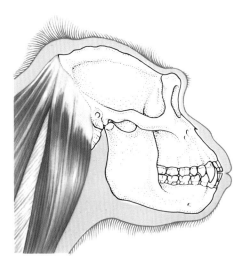

Compare these two australopithecine skulls with the human and gorilla skulls. Notice where the skull was joined to the backbone. And look at the marks where the neck muscles joined the skull. Do you agree that the australopithecine skulls are more like the human skull than the gorilla skull?

This is one piece of evidence that the australopithecines walked upright. We can find further evidence in other parts of the skeleton.

australopithecine skull from Olduvai Gorge, Tanzania

Human beings walk upright. Their heads are well balanced on top of their backbones. The backbone meets the skull towards the middle. The neck muscles that support the head are small and attached quite low down on the skull.

The backbone of a gorilla meets the skull towards the back, and the head needs much more support. The neck muscles are large and strong and are attached fairly high up on the skull.

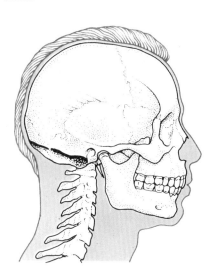

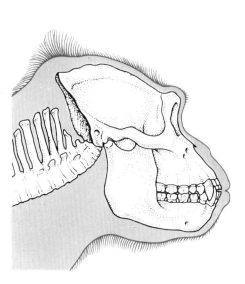

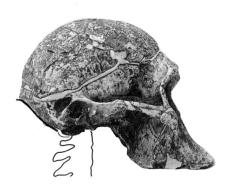

australopithecine skull from Sterkfontein, South Africa

The hip

human being

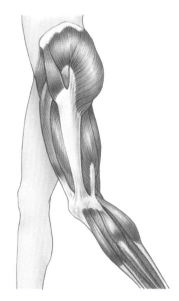

gorilla

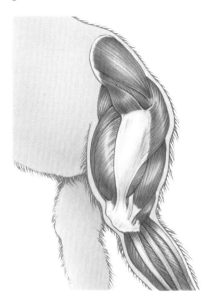

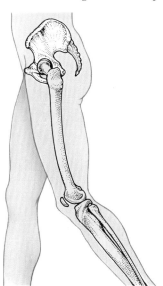

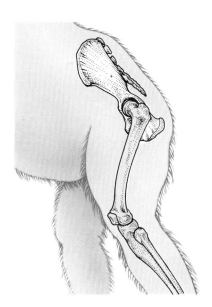

The hip is important in walking because most of the muscles used in walking are attached to it. The relative size of some of these walking muscles is quite

different in human beings and gorillas. And the shape of the hip bone reflects these differences.

Look at this australopithecine hip bone. You can see that it is short and broad like a human hip bone, not long and narrow like a gorilla hip bone.

australopithecine hip bone from Sterkfontein, South Africa

43

The thigh bone

The thigh bone is also important in walking. Again, many of the muscles used in walking are attached to it. Compare this australopithecine thigh bone with the human and gorilla thigh bones. Even though the australopithecine thigh bone is incomplete, you can see that it is much more like the human thigh bone – it has a long neck, and the shaft is long and slightly curved.

The foot

Foot bones are small and fragile, and are not often preserved as fossils. But a few australopithecine foot bones have been found. The bones from several different individuals have been used to reconstruct an australopithecine foot.

Compare this possible australopithecine foot with the feet of a human being and a gorilla. Notice that the toes are short like ours, and the big toe points forward. The gorilla's toes are very long, and its big toe sticks out at the side.

So the australopithecines seem to have had feet more like ours than like gorillas'.

What kind of footprints do you think they made?

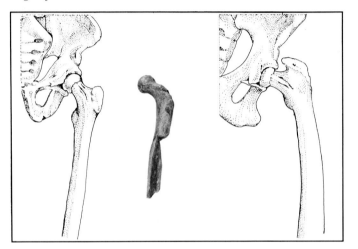

human being

australopithecine thigh bone from Koobi Fora, Kenya

gorilla

 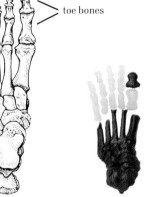 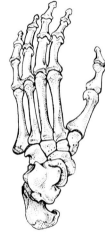

toe bones

human being

possible australopithecine foot, based on foot and big toe bones from Olduvai Gorge, Tanzania

gorilla

Lucy

In 1974, the remains of a skeleton about 3 million years old were found at Hadar, in the Afar region of Ethiopia. Nearly half the skeleton had been preserved.

This skeleton, which is probably female, was nicknamed 'Lucy'. 'Lucy' has since been named as a new kind of australopithecine, *Australopithecus afarensis* or the **Hadar australopithecine.**

Look carefully at this photograph of 'Lucy'. See if you can decide whether or not she walked upright like a human being. Look especially at the hip bone and the thigh bone.

All the evidence from the different parts of the skeleton suggests that 'Lucy' – and all the other australopithecines – must have walked upright, like human beings.

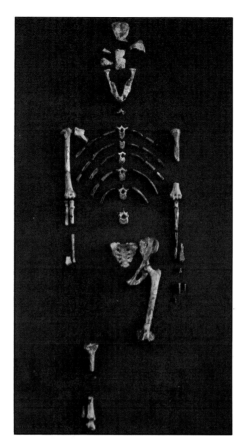

'Lucy', incomplete skeleton of an australopithecine from Hadar, Ethiopia

But the australopithecines may not have walked in quite the same way as human beings. The blade of an australopithecine hip bone seems to have been more at the back of the body than at the side – in a human being it is at the side. And the neck of an australopithecine thigh bone is very flat, not rounded like the neck of a human thigh bone. These two differences suggest that the australopithecines probably did not stride along in the same way that we do.

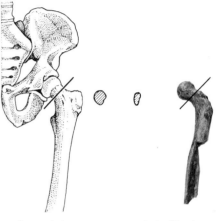

human being   australopithecine

Fossilized footprints found at Laetoli, Tanzania.
Made 3.75 million years ago by creatures that walked on two legs – Hadar australopithecines?

Different kinds of australopithecines

In addition to the Hadar australopithecines, the remains of two other kinds are known.

Robust australopithecines
Australopithecus robustus and *boisei*

These australopithecine skulls and jaws belong to a group known as the **robust australopithecines.**

Look at these teeth. The robust australopithecines have very large chewing teeth, but small canines and biting teeth. They were probably vegetarians, and their teeth are specially adapted to their vegetarian diet.

Gracile australopithecines
Australopithecine africanus

Another group of australopithecines – the **gracile** or **slender australopithecines** – seem to have few unique characteristics of their own.

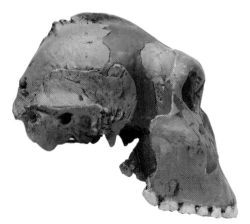

Part of the skull of a robust australopithecine from Olduvai Gorge, Tanzania

Part of the skull of a robust australopithecine from Swartkrans, South Africa

Part of the skull of a gracile australopithecine from Sterkfontein, South Africa

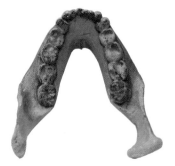

Jaw of a robust australopithecine from Peninj, Tanzania

Jaw of a robust australopithecine from Swartkrans, South Africa

Jaw of a gracile australopithecine from Sterkfontein, South Africa

Are the Hadar australopithecines more closely related to man than the other australopithecines are?

The robust australopithecines show the same pattern of tooth eruption as human beings, and the base of their skull is a similar shape to ours.

The gracile australopithecines have a domed sloping forehead over the browbridge, similar to that of modern humans.

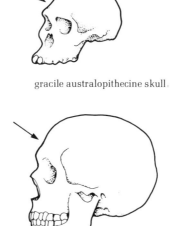

gracile australopithecine skull

The Hadar australopithecines are the only kind of australopithecines with cheekbones that are level with the 1st molar teeth – as they are in human beings.

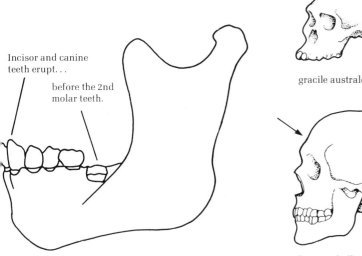

Incisor and canine teeth erupt...

before the 2nd molar teeth.

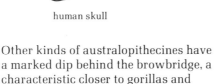

human skull

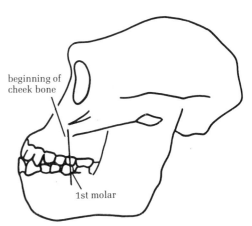

beginning of cheek bone

1st molar

However, the robust australopithecines have specialized teeth, a characteristic that we do not share.

Other kinds of australopithecines have a marked dip behind the browbridge, a characteristic closer to gorillas and chimpanzees than to human beings.

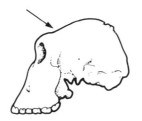

robust australopithecine skull

They also show many unspecialized characteristics, such as smaller premolar teeth, found in human beings but not in the other australopithecines.

Each kind of australopithecine has special features that could link it with human beings.

On the evidence, we cannot decide which of these australopithecines is our closest relative – we must wait for new evidence.

Chapter 5

Man makes tools

The habilines

Another group of creatures thought to be our fossil relatives lived in East Africa from about 2 million to 1.5 million years ago. They have been called the **habilines**, and they were probably the first creatures to make tools.

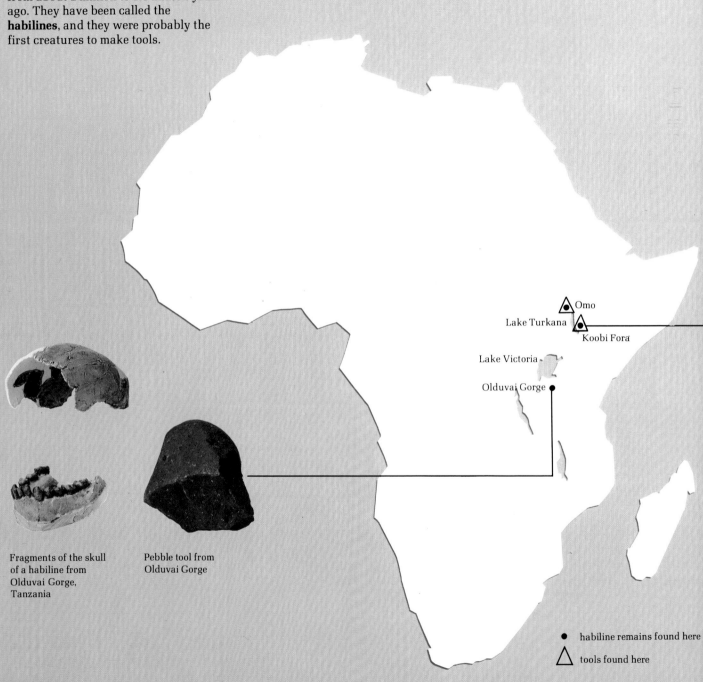

Omo

Lake Turkana

Koobi Fora

Lake Victoria

Olduvai Gorge

Fragments of the skull of a habiline from Olduvai Gorge, Tanzania

Pebble tool from Olduvai Gorge

● habiline remains found here

△ tools found here

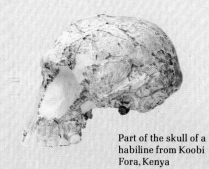

Part of the skull of a habiline from Koobi Fora, Kenya

Flake tool and pebble tool from Koobi Fora

How are the habilines related to man?

The habilines are closely related to human beings and the australopithecines. They share the same important characteristics:

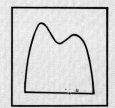

● 2 cusps on premolars

● walking on 2 legs

If you look closely at this skull you can see evidence that the habilines walked on two legs.

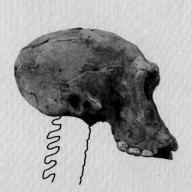

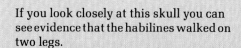

Part of the skull of a habiline from Olduvai Gorge, Tanzania

Are the habilines more closely related to the australopithecines?

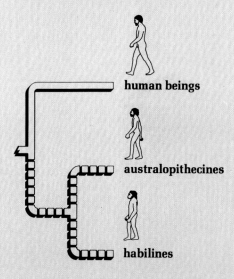

human beings

australopithecines

habilines

Or are they more closely related to human beings?

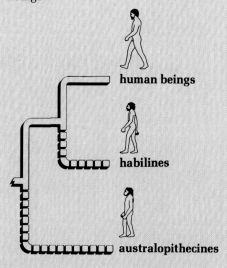

human beings

habilines

australopithecines

We think that the habilines are more closely related to human beings than to the australopithecines because there is evidence that they had large brains, and they made tools.

Man has a large brain

A large brain is another important human characteristic. Of course, there are other, much larger animals that have even bigger brains. But no other animals of the same size have such a large brain.

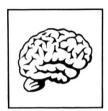

For example, a 60 kilogram chimpanzee has a much smaller brain than a 60 kilogram human being has.

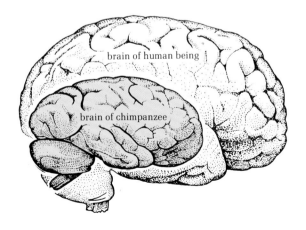

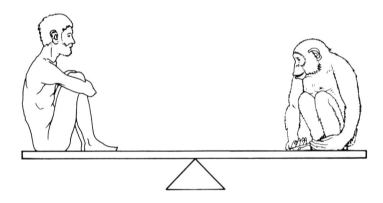

brain of human being

brain of chimpanzee

How do we know the brain size of a fossil?

Of course, we cannot measure the brains of fossils directly, because they are never preserved. But we can get some idea of how big the brain of a fossil was if we make a cast of the inside of its skull, or even part of its skull.

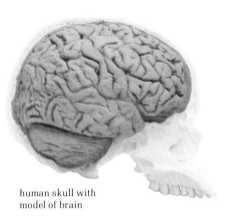

human skull with
model of brain

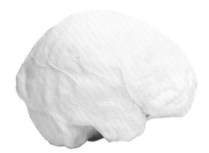

cast of the inside
of a human skull

Such a 'brain cast' is very slightly larger than the brain itself, because there is a space between the skull and the brain.

The habilines had large brains

Although the habiline brain cast is much smaller than the brain of a human being, the habilines were much smaller creatures. From the size of their bones, we can estimate that they probably weighed only about 40 kilograms – about the same as a twelve-year-old child. So, for their size, the habilines had fairly large brains.

Of course, the australopithecines were also small. And, for their size, they also had quite large brains. But we think that the difference between the habiline and the australopithecine brain sizes is important because the habilines made tools. (You will find out more about this later in the chapter.)

Man makes tools

Human beings are the only living creatures that manufacture tools – we *use* tools to *make* tools.

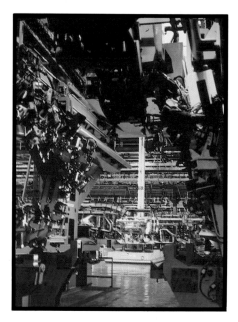

Other animals including some birds, chimpanzees, and possibly some australopithecines, may select objects to use as tools. They may even modify the objects to make them into tools. But they have never been known to use tools to make tools, as we do.

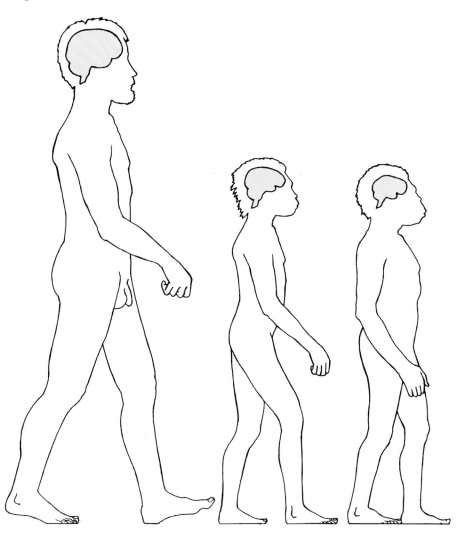

human being   habiline   australopithecine

Making tools is very different from merely using them. It involves solving problems...

No other living creature shares man's tool-making behaviour — it is a unique human characteristic.

The habilines made tools

The habilines were probably the first creatures to make tools. There is evidence that they collected stones, often from many kilometres away, and used other stones to reshape them into tools.

They made different tools for different purposes. Their 'toolkit' includes flakes and knife-like tools, as well as a variety of choppers and scrapers.

We can guess that the australopithecines may also have used stones and bones as tools, but there is no evidence that they made tools as the habilines did.

The remains of a habiline camp at Olduvai Gorge, Tanzania

The habiline 'toolkit' found at Olduvai Gorge is usually referred to as the **Oldowan industry.** It includes:

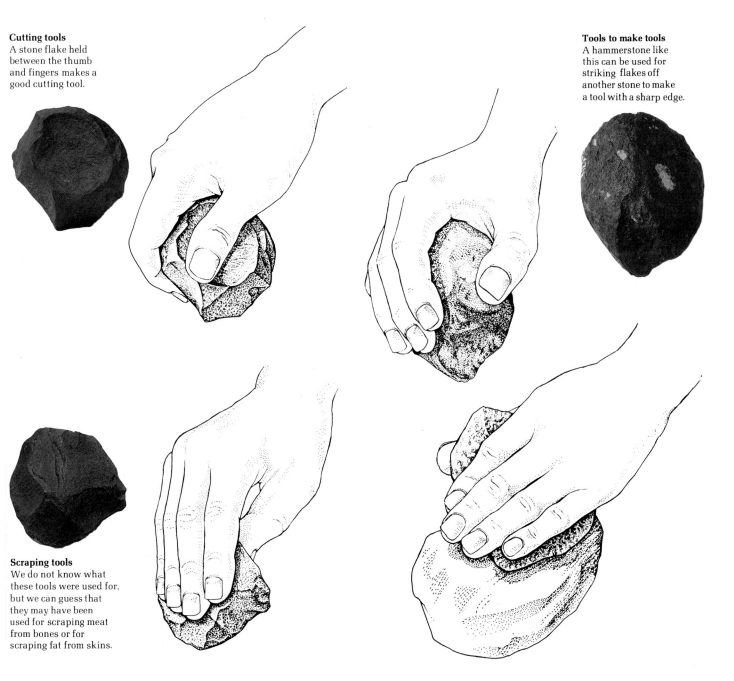

Cutting tools
A stone flake held between the thumb and fingers makes a good cutting tool.

Tools to make tools
A hammerstone like this can be used for striking flakes off another stone to make a tool with a sharp edge.

Scraping tools
We do not know what these tools were used for, but we can guess that they may have been used for scraping meat from bones or for scraping fat from skins.

The habiline way of life

After watching hyenas bring down an antelope, the habilines would chase them away and steal the kill for themselves.

Some of the habilines made tools with special
stones they carried for the purpose. The tools were
then used to butcher the carcass.

Having removed all the meat they could from the carcass, the habilines moved on, leaving their tools behind.

Were the habilines human?

Yes. Toolmaking is such an important human characteristic that we can think of the habilines as human beings.

Obviously toolmaking is very different from the other characteristics that we have used to test relationships. It is not a physical characteristic at all. It is a characteristic of behaviour, and it can be learned.

But it takes a great deal of intelligence to solve problems by making tools, and this is why we think that toolmaking is so important. The habilines' varied toolkit is good evidence that they had quite clever brains.

The scientific name for all living human beings is *Homo*, which is Latin for 'man'. We have given this name to the habilines too.

But the habilines are so different from us that we can think of them as a different kind of human being. So we have given them their own scientific name – *Homo habilis*, which means 'handy man'.

Modern human beings are called *Homo sapiens*, which means 'wise man'. In order to distinguish them from fossil human beings like the habilines, we shall refer to modern human beings as **modern humans** in the rest of the book.

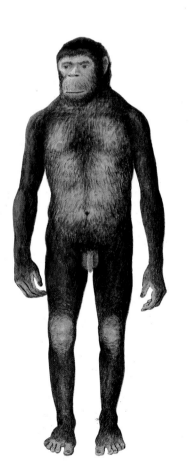

Habiline
Homo habilis

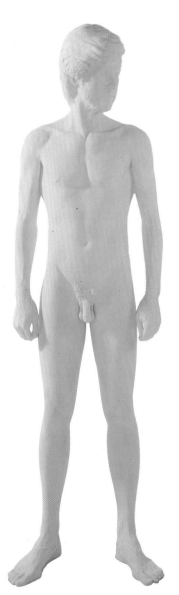

Modern human
Homo sapiens

Chapter 6
Man uses fire

The *Homo erectus* people

Remains of other fossil human beings have been found in Africa and South East Asia in rocks ½ - 1½ million years old. These fossils have been given the name *Homo erectus* – 'upright man'.

Later *Homo erectus* people spread to Northern Asia and to Europe. There is evidence that these people used fire – which would certainly have helped to solve the problem of how to keep warm in these cooler northern climates.

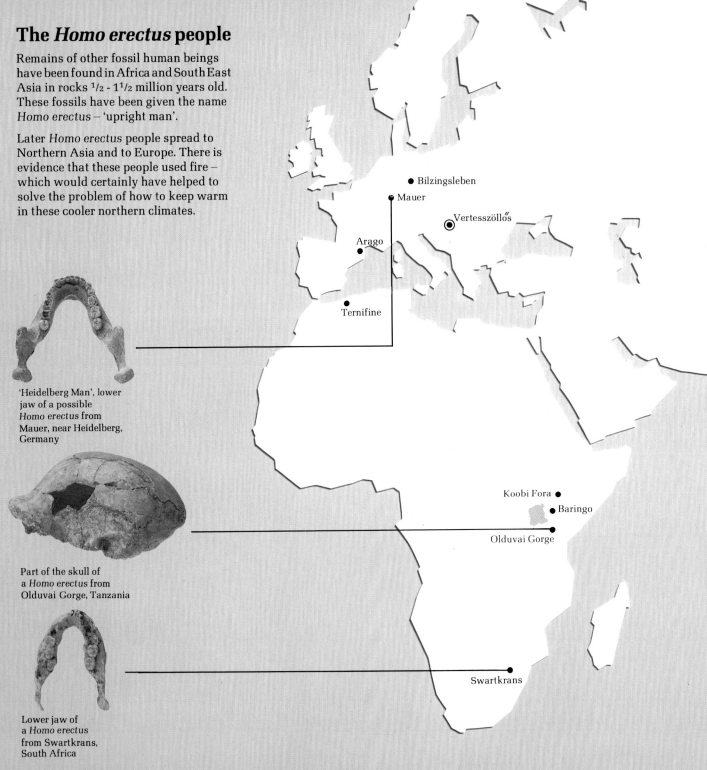

● Bilzingsleben

● Mauer

◉ Vertesszöllős

● Arago

● Ternifine

Koobi Fora ●

● Baringo

Olduvai Gorge ●

● Swartkrans

'Heidelberg Man', lower jaw of a possible *Homo erectus* from Mauer, near Heidelberg, Germany

Part of the skull of a *Homo erectus* from Olduvai Gorge, Tanzania

Lower jaw of a *Homo erectus* from Swartkrans, South Africa

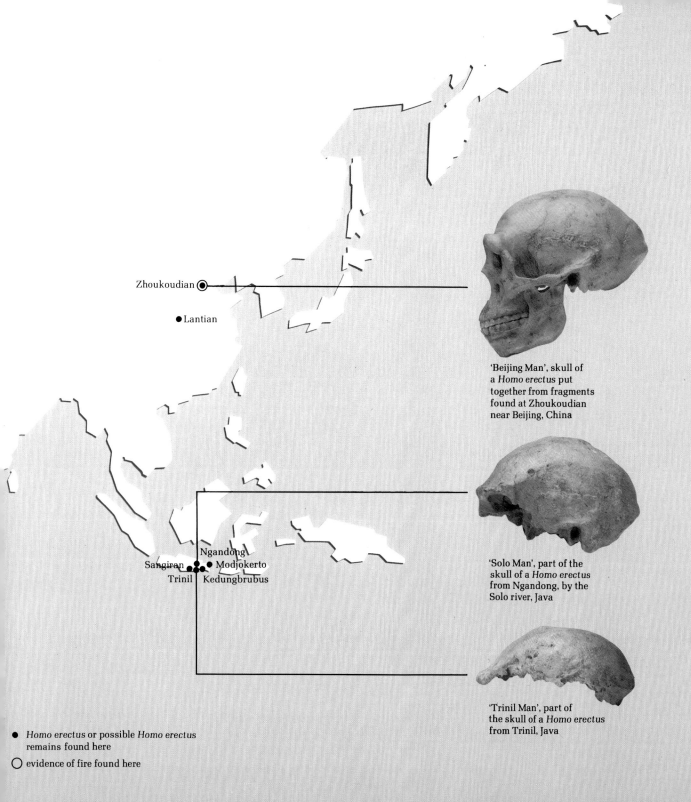

Zhoukoudian ◉

● Lantian

Ngandong
Sangiran ● ● Modjokerto
Trinil ● Kedungbrubus

'Beijing Man', skull of
a *Homo erectus* put
together from fragments
found at Zhoukoudian
near Beijing, China

'Solo Man', part of the
skull of a *Homo erectus*
from Ngandong, by the
Solo river, Java

'Trinil Man', part of
the skull of a *Homo erectus*
from Trinil, Java

● *Homo erectus* or possible *Homo erectus*
remains found here
○ evidence of fire found here

How are the *Homo erectus* people related to modern humans?

If the *Homo erectus* people are closely related to modern humans and the habilines, they must share the same characteristics...

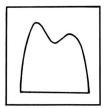

● 2 cusps on premolars

● walking on 2 legs

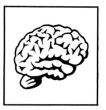

● a large brain

● tool-making

You can find evidence for some of these characteristics in these *Homo erectus* remains found at Trinil in Java...

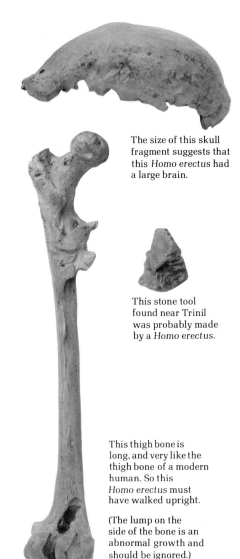

The size of this skull fragment suggests that this *Homo erectus* had a large brain.

This stone tool found near Trinil was probably made by a *Homo erectus*.

This thigh bone is long, and very like the thigh bone of a modern human. So this *Homo erectus* must have walked upright.

(The lump on the side of the bone is an abnormal growth and should be ignored.)

So the *Homo erectus* people must be closely related to modern man and the habilines. But which are they *more* closely related to – modern humans or the habilines?

Like modern humans, the *Homo erectus* people used fire. And, for their size, they had larger brains than the habilines.

brain cast of a *Homo erectus*

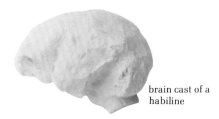

brain cast of a habiline

They also had smaller teeth – their wisdom teeth in particular were small and much like our own. All this suggests that the *Homo erectus* people are more closely related to modern humans than to the habilines...

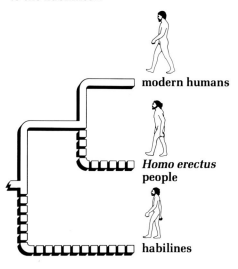

modern humans

Homo erectus people

habilines

64

Man uses fire

Fire using is an important characteristic of modern humans. We use fire to solve many different kinds of problems.

Using fire to solve problems is similar to making tools to solve problems. It requires intelligence and imagination, and can be seen as evidence of a fairly advanced brain. No other living creature uses fire – it is a unique characteristic of humans.

We use fire for. . .

heating

cooking

making things

The *Homo erectus* people used fire

There is evidence that the *Homo erectus* people who lived at Zhoukoudian near Beijing in China were able to use fire. Ash layers, charcoal and burnt bones have been found with *Homo erectus* remains in a cave site at Zhoukoudian.

Evidence for fire at Zhoukoudian
Reconstruction based on information from the excavations at Zhoukoudian

The excavation site at Zhoukoudian

Ash layer
– from burnt-out fires

Charcoal
– the remains of burnt-out fires

Scorched stone
– hot stones may have been plunged into water to heat it for cooking purposes

Burnt bone
– from a deer that may have been killed and eaten

Shells of hackberry nuts
– from nuts that may have been crushed and eaten

The *Homo erectus* way of life

We have evidence that the *Homo erectus* people used fire, and we can imagine how this may have influenced their way of life. . .

A fire would provide a centre for a permanent home base. People would gather round the fire, and so would tend to live in groups.

Some members of the group would band together to go and hunt large animals. Others would gather seeds and fruits. Food would be brought back to the home base to be cooked and shared out amongst the members of the group.

Archaeologists are looking for evidence of such a way of life at the various sites where the remains of *Homo erectus* people, or their tools, have been found.

Not our direct ancestors

The *Homo erectus* people were not quite like us. Compare this reconstructed *Homo erectus* skull with the modern human skull. You will see that the *Homo erectus* skull has several characteristics that the modern skull does not share.

Because of these special characteristics, we think that some, at least, of the *Homo erectus* people were not our direct ancestors.

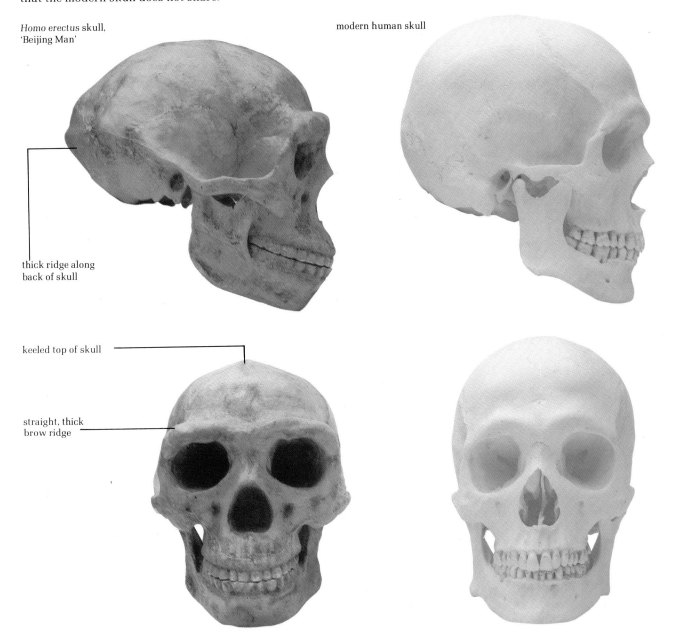

Homo erectus skull, 'Beijing Man'

modern human skull

thick ridge along back of skull

keeled top of skull

straight, thick brow ridge

What shall we call this fossil?

This fossil skull from Petralona in Greece shows only one of the *Homo erectus* characteristics – a thick ridge along the back of the skull. Experts cannot agree whether it is a *Homo erectus* skull with some *Homo sapiens* characteristics, or, a *Homo sapiens* skull with some *Homo erectus* characteristics.

Names like *Homo erectus* are useful, because they help us to talk about *groups* of fossils, rather than about individual specimens. But they are only labels, and can cause problems when a fossil – like the Petralona skull – shows characteristics of more than one group.

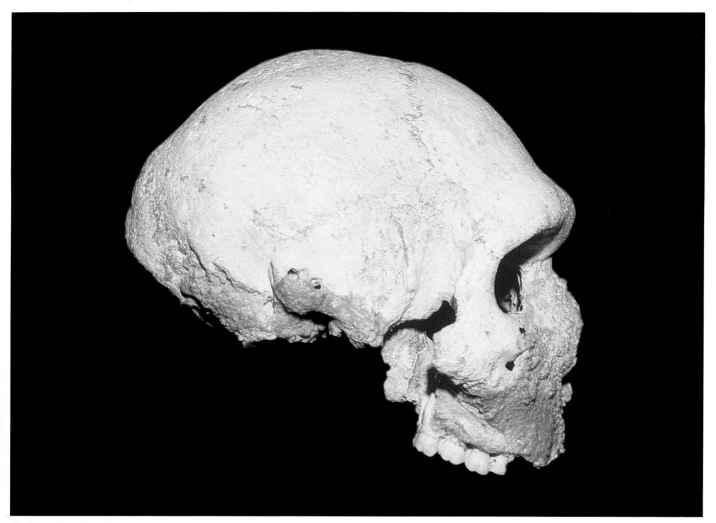

Skull from Petralona, Greece

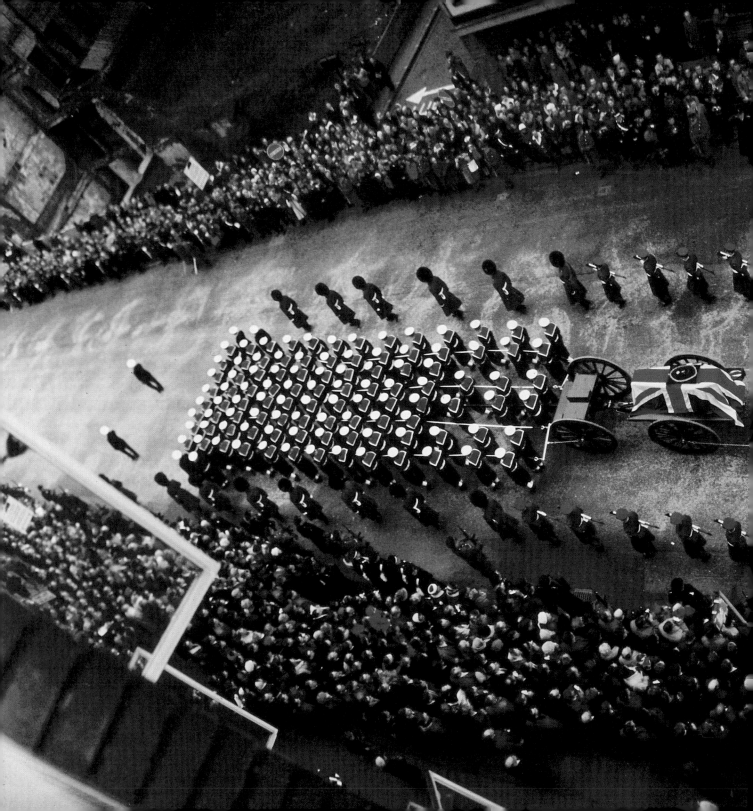

Chapter 7

Man has ceremonies

The neandertals

The neandertals lived in Europe and the Middle East before and during the last Ice Age. Most of them lived between 100 000 and 40 000 years ago. They were probably the first people to have ceremonies for their dead.

They were given the name **neandertal** (sometimes spelt neanderthal) because the first specimen to be scientifically described was found in a cave in the Neander Valley – *Neander Thal* in Old German.

Part of the skull of a neandertal from the Neander Valley, Germany – the first specimen to be given the name neandertal

Part of the skull of a neandertal found at Forbes Quarry, Gibraltar in 1848 – the second neandertal to be found

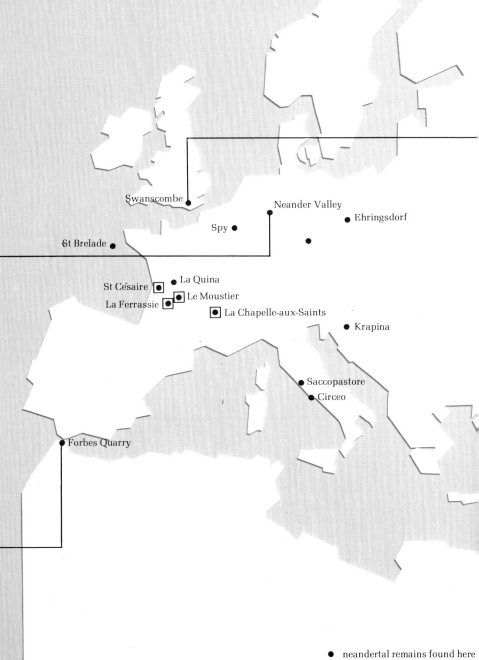

Swanscombe

Neander Valley

Ehringsdorf

Spy

St Brelade

St Césaire

La Quina

Le Moustier

La Ferrassie

La Chapelle-aux-Saints

Krapina

Saccopastore

Circeo

Forbes Quarry

● neandertal remains found here

□ evidence of burial found here

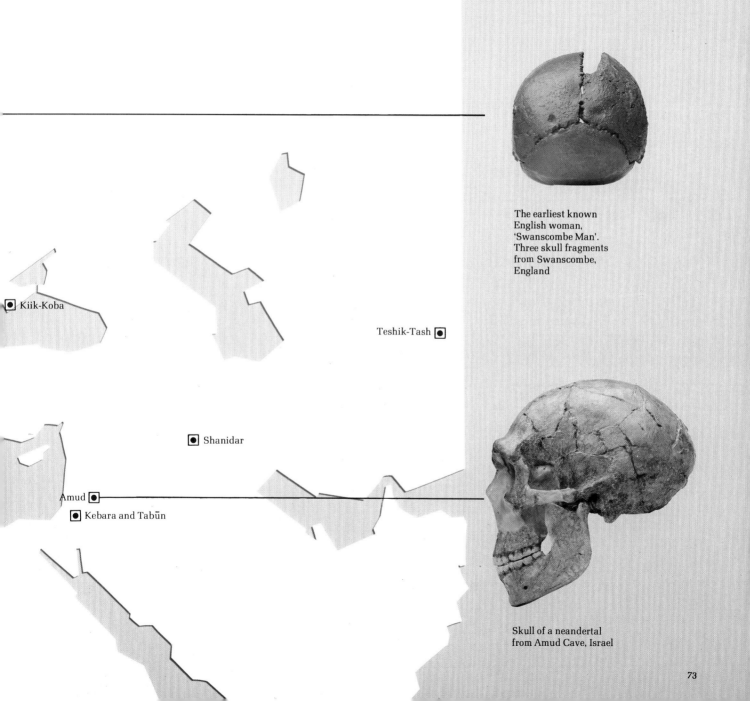

The earliest known
English woman,
'Swanscombe Man'.
Three skull fragments
from Swanscombe,
England

● Kiik-Koba

Teshik-Tash ●

● Shanidar

Amud ●

● Kebara and Tabūn

Skull of a neandertal
from Amud Cave, Israel

How are the neandertals related to modern humans?

A neandertal skeleton was discovered during excavations at Tabūn in Israel. By the skeleton, flake tools and burnt bones were also found – providing evidence that the neandertals made tools and used fire.

Flake tool found with neandertal skeleton at Tabūn

Burnt bone found with neandertal skeleton at Tabūn

This neandertal woman is based on the skeleton found at Tabūn. (At the end of the chapter, you can find out how she was made.) You can see that she walked upright. And the size of her head suggests that she had a large brain.

So the neandertals must be closely related to modern humans and the *Homo erectus* people, because they share the same characteristics. But which are they *more* closely related to – modern humans or the *Homo erectus* people?

The neandertals have large brow ridges, like the *Homo erectus* people have.

Homo erectus skull, 'Beijing Man'

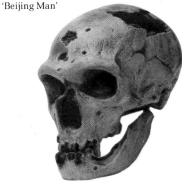

neandertal skull from La Chapelle-aux-Saints, France

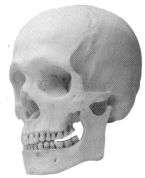

modern human skull

So they could be closely related to the *Homo erectus* people. . .

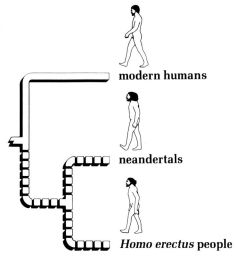

modern humans

neandertals

Homo erectus **people**

But like modern humans, the neandertals had very large brains – most of them had a brain larger than 1300 ml. And, like modern humans, they buried their dead. So we think that the neandertals are more closely related to modern humans. . .

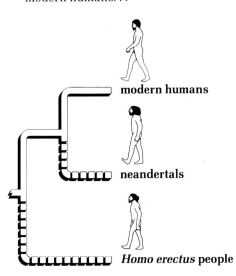

modern humans

neandertals

Homo erectus **people**

Homo sapiens

Compare this neandertal skull with a modern human skull. The most obvious difference is in the general shape of the face. But there are other differences. . .

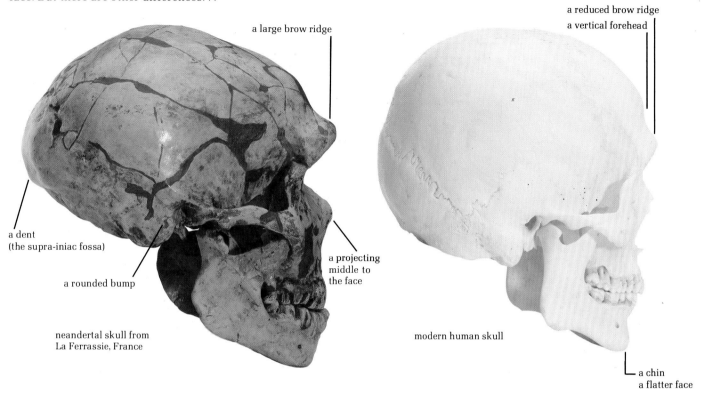

a large brow ridge

a reduced brow ridge
a vertical forehead

a dent
(the supra-iniac fossa)

a rounded bump

a projecting middle to the face

neandertal skull from
La Ferrassie, France

modern human skull

a chin
a flatter face

There are also differences in other parts of the skeleton. But because the neandertals and modern humans share these two important characteristics. . .

- an average brain size of 1400 ml
- burial of the dead – ceremonies

. . .the differences are often considered unimportant and the neandertals and modern humans are grouped together in the same species – *Homo sapiens*.

The neandertals' unique characteristics are often recognized by calling them a subspecies of modern humans – *Homo sapiens neanderthalensis*. If so, they were an ancient form of *Homo sapiens* that died out about 30 000 years ago.

Man has ceremonies

Ceremonies are an important characteristic of modern humans. We live in societies, and we recognize ourselves as part of a society when we take part in its ceremonies.

We have ceremonies. . .

Ceremonies like these reflect our complex social structure, and are an important part of human culture. Like tool-making and the use of fire, ceremonies provide indirect evidence of our intellectual and cultural development.

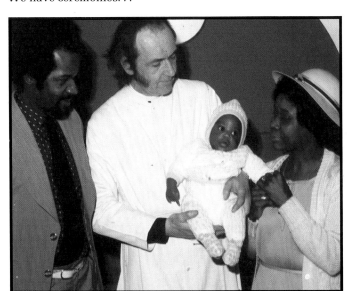

when a baby is born

when two people are married

when someone dies

The neandertals had ceremonies

The neandertals were probably the first people who had ceremonies for their dead. There is evidence for such ceremonies at several sites where neandertal remains have been found.

These are the remains of a neandertal child found buried at Teshik-Tash in the USSR. The body was placed in a shallow pit, and pairs of goat horns were placed in a circle around it. A fire was lit beside the grave.

Reconstruction of a neandertal burial, based on remains found at Teshik-Tash, USSR

A neandertal from Kent?

These are the remains of the earliest known English woman – 'Swanscombe Man'. They are perhaps 250 000 years old and were found at Swanscombe in Kent, not far from London.

The three skull bones were discovered at separate times (in 1935, 1936 and 1955). But they clearly come from the same person – a young woman.

On the back of her skull, there is the small dent (the supra-iniac fossa) that is one of the special characteristics of the neandertals. So the Swanscombe people may have been some of the earliest neandertals.

Other evidence of these ancient inhabitants of Britain has also been found in the river gravels at Swanscombe. . .

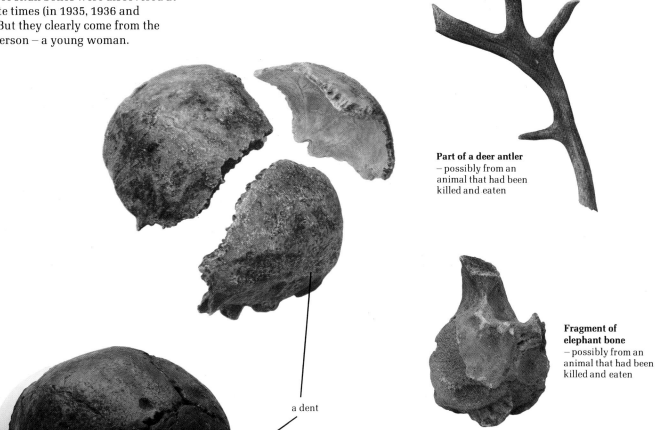

a dent

Part of a deer antler
– possibly from an animal that had been killed and eaten

Fragment of elephant bone
– possibly from an animal that had been killed and eaten

Flint hand axe
Hand axes like this have a long cutting edge, and make better cutting tools than stone flakes do.

From these remains, we can begin to guess at how the Swanscombe people lived.

In the Kent area at the time there were several large animals that could have been sources of food. The Swanscombe people may have hunted a kind of fallow deer, and occasionally butchered elephants and rhinoceroses found trapped in mud or marshy ground. The meat was probably then shared out.

Reconstructing a neandertal

For the exhibition of **Man's place in evolution**, one of the Natural History Museum's modelmakers produced a unique reconstruction of a neandertal woman. He based her on the 50 000 year old remains of a female neandertal skeleton, found at Tabūn, near Mount Carmel in Israel.

The neandertal woman took about three months to reconstruct. First of all, the modelmaker used measurements of the Tabūn bones to work out her height, the relative lengths of her limbs, and so on. Then, using his own

knowledge of anatomy, together with published information about neandertals, and advice from Museum scientists, he produced drawings to show what the reconstruction would look like.

When the drawings had been approved, and the model's exact pose had been decided on, the modelmaker could begin to plan the armature – the supporting framework on which the model would be built. He made the armature from wood and steel, and added a plaster cast of the original skull from Tabūn.

Polystyrene was used to form the bulk of the woman's torso, and the rest of the body was built up from layers of scrim (sacking) soaked in plaster. Later, a layer of plasticene was added. The modelmaker used the plasticene to create all the fine details – the exact contours of the woman's flesh, her skin texture, hair and so on.

Research and drawing

Welding the armature

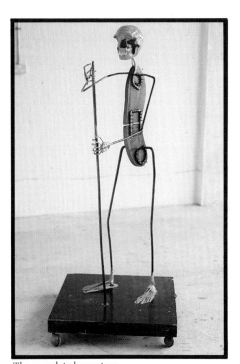

The completed armature

Building up the model with scrim and plaster

Modelling the details in plasticene

The completed 'plasticene' model

When the model had been completed to everyone's satisfaction, the modelmaker made a mould from it. Because of the model's complex shape, the mould had to be carefully planned, and was made in seven separate sections. Each of these was built up on the model from layers of liquid silicon rubber, strengthened with pieces of cloth. Each section was then encased in a strong 'jacket' of glass-reinforced plastic (GRP). When all the sections were complete, they were bolted together around the model to ensure a perfect fit.

The mould was then taken apart and the inside of each section was painted with a 'gel coat' of polyester resin. Specially-made glass eyes were put in place, and the mould was lined with layers of GRP. While this was still wet, the pieces were bolted back together again.

When the GRP was dry, the mould was taken apart. Inside, the GRP had formed a strong, hollow, lightweight replica or cast of the original model.

This cast was carefully cleaned and all the 'seam lines' were removed. Finally, the neandertal woman was painted, ready for display.

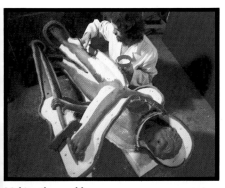
Making the mould

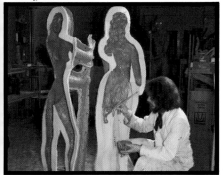
Applying the 'gel coat' to the mould

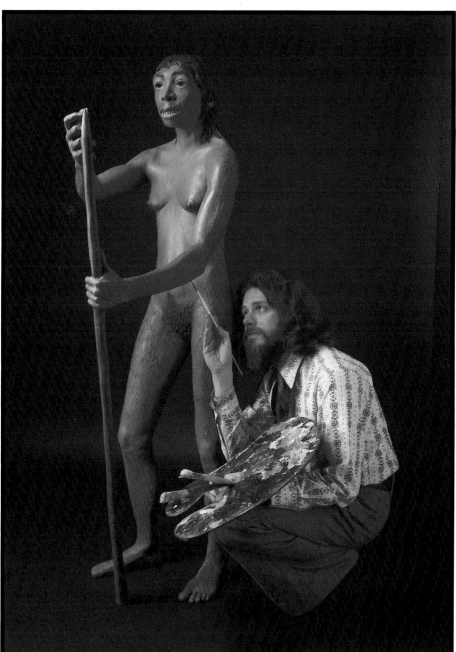
The finishing touches

Chapter 8
Man uses symbols

Modern humans

By about 30 000 years ago modern humans had spread to nearly all parts of the world. Physically, they were just like people living today.

Man the hunter

These early people lived mainly by hunting large animals such as deer, horses and mammoths. In order to kill these animals, they must have lived and hunted in large organized groups.

Each group would probably have had a leader. And the work of hunting, making traps, building shelters and caring for children would have been divided amongst members of the group.

This kind of social organization could best be achieved if the members of the group could communicate with each other using language.

Can we find any evidence that they did use language?

Modern humans use symbols

Language is an important characteristic of modern humans. In language, we use sounds – words – to represent things and ideas. In a similar way we use many other kinds of symbols to represent things, ideas and feelings.

We use symbols in. . .

language

music

art

Man the hunter used symbols

The hunters of the last Ice Age used symbols too. They represented the animals they hunted in magnificent cave paintings. They carved female figures out of stone and bone, or moulded them in clay. They even made simple musical instruments.

There is no way of knowing when humans first began to use language. We cannot tell from the fossils. But we can be fairly sure that the people who made these paintings and carvings also used language.

Art and language are closely linked. They both involve the use of symbols to represent things and ideas. This is why we think that the Ice Age hunters must have used language – their art is our evidence.

gesture

mathematics

No other creature uses this wide range of symbols. The use of symbols in art, music, language and so on is a unique characteristic of modern humans.

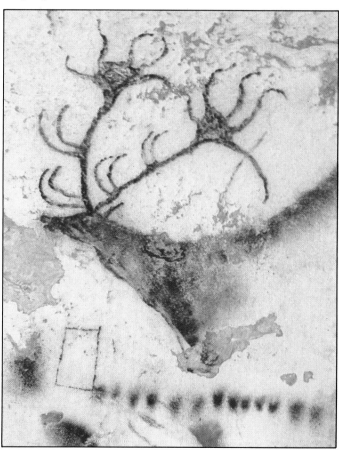

Cave paintings, from Lascaux in France. About 17 000 years old. (Original about 1.4 metres wide)

Woman carved in soapstone, from Balzi Rossi in Italy. About 20 000 years old. (Actual size)

Woman's head carved in ivory, from Brassempouy in France. About 18 000 years old. (Actual size)

The development of farming

Towards the end of the last Ice Age, people began to farm in several different parts of the world. And, by about 7000 years ago, farmers already grew crops such as wheat and barley, and kept animals such as sheep, cattle and pigs.

Farming probably developed in response to a shortage of food. Towards the end of the Ice Age, the human population was steadily increasing. But climatic changes had led to the extinction or migration of many of the larger animals. So, in many areas, there were far fewer animals to hunt. Also, in dry areas, people would be forced to live near permanent supplies of water.

We can imagine that the people in these settled communities must have been forced to store food for hard times. They probably stored nourishing grains, and kept animals tethered as a kind of 'walking food store'. Such practices could gradually have developed into farming.

Maize was first cultivated in Central America.

Tehua án Valley ● ● Guilá Naquitz Cave

Llamas were first domesticated in South America.

Uchcumachay Cave ●

◉ Ayacucho Basin

Evidence of farming more than 7000 years ago:

● cultivated plants
○ domesticated animals
⌒ Fertile Crescent

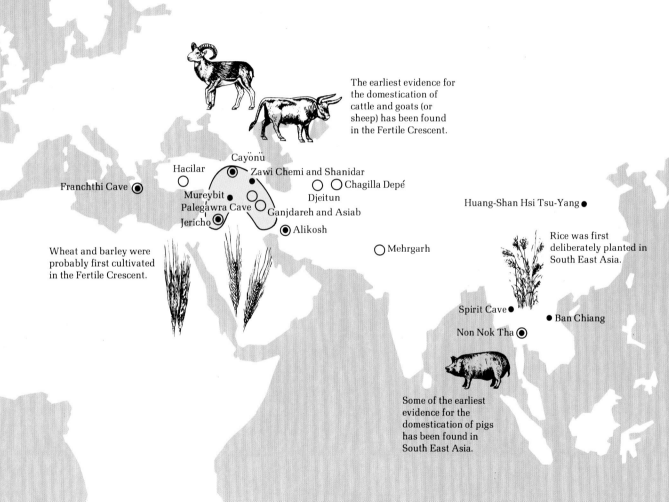

The earliest evidence for the domestication of cattle and goats (or sheep) has been found in the Fertile Crescent.

Cayönü

Hacilar

Zawi Chemi and Shanidar

Franchthi Cave

Chagilla Depé

Djeitun

Huang-Shan Hsi Tsu-Yang

Mureybit

Palegawra Cave

Ganjdareh and Asiab

Jericho

Alikosh

Mehrgarh

Rice was first deliberately planted in South East Asia.

Wheat and barley were probably first cultivated in the Fertile Crescent.

Spirit Cave

Ban Chiang

Non Nok Tha

Some of the earliest evidence for the domestication of pigs has been found in South East Asia.

Humans farm

Farming is another important characteristic of modern humans. Farming involves. . .

storing crops

planting crops

harvesting crops

herding animals

feeding animals

breeding animals

Farmers must be able to plan for the future. They must store grain for planting next year, and store food to feed themselves and their animals during the winter.

Farming does not require any new mental or physical skills. The Ice Age hunters would have needed the same amount for foresight, imagination and planning when they were hunting big game.

But farming *was* a very important step forward for modern humans, because it was a solution to the problem of getting enough to eat. Farming produces more food from the same amount of land. So the communities that adopted farming would have had an advantage over the hunters. Their populations would have been able to increase, and these early farming settlements were the beginning of the towns and cities of the modern world.

. . . so that people can eat.

Evidence of early farming

There is evidence that the people who lived at Jericho, near the Dead Sea, 9000 years ago were farmers.

Excavations show that there was a large settlement at Jericho – as many as 2000 people may have lived there. Such a large population could not have been supported by hunting alone.

Part of a house excavated at Jericho

There is evidence for cultivated barley at Jericho. Grains of 2-row barley found there are larger than grains of wild 2-row barley, and are believed to come from cultivated plants.

'cultivated' two-row barley

wild two-row barley

This flint sickle from Jericho was evidently used for cutting ears of corn – its edge has been polished by the silica in their stalks. Sickles like this would be needed only where there was a lot of corn to cut.

reconstructed flint sickle

There is also evidence for domesticated goats. Early domesticated animals were usually tethered and sometimes poorly fed. Small-sized animals were better able to survive this treatment. So, at first, domesticated animals were smaller than wild animals of the same species.

bone of 'domesticated' goat from Jericho

bone of wild goat from Jericho

This quern (grindstone) from Jericho must have been used for grinding corn. Grindstones very like this are still in use in some parts of the world.

When we compare ourselves with our fossil relatives, we can find evidence that humans have evolved.

When we look at the fossil remains, we find evidence of physical differences. For example we are taller than the first human beings – the habilines. And we have much larger brains.

There is also evidence that human behaviour has become much more complicated. For example the *Homo erectus* people used fire and, later, the neandertals took part in ceremonies for their dead.

The first modern people lived in organized groups and hunted large animals, which they depicted in their paintings. Some of their descendants lived in settled communities that developed first into farming villages, and then into the towns and cities of the modern world.

From the time of the first toolmakers, humans have survived by controlling the environment in a way that no other animal has.

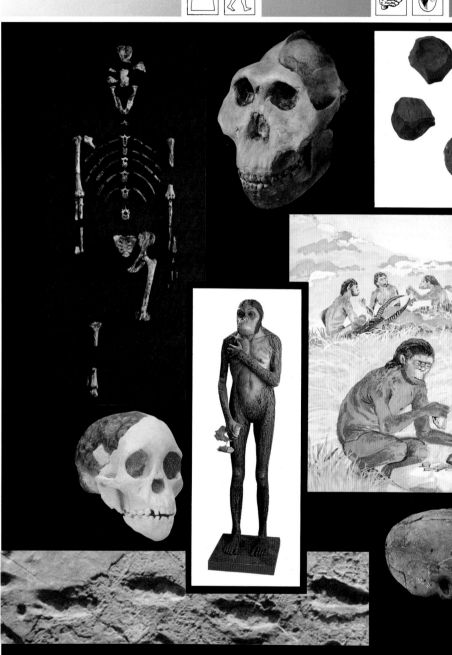

1¹/₂ million years ago

100 000 years ago

30 000 years ago.

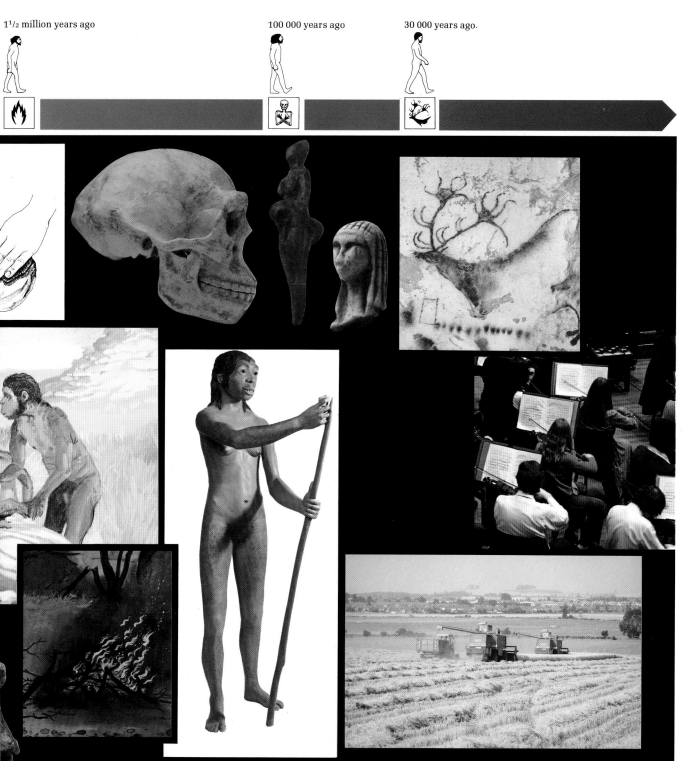

Man's place in evolution

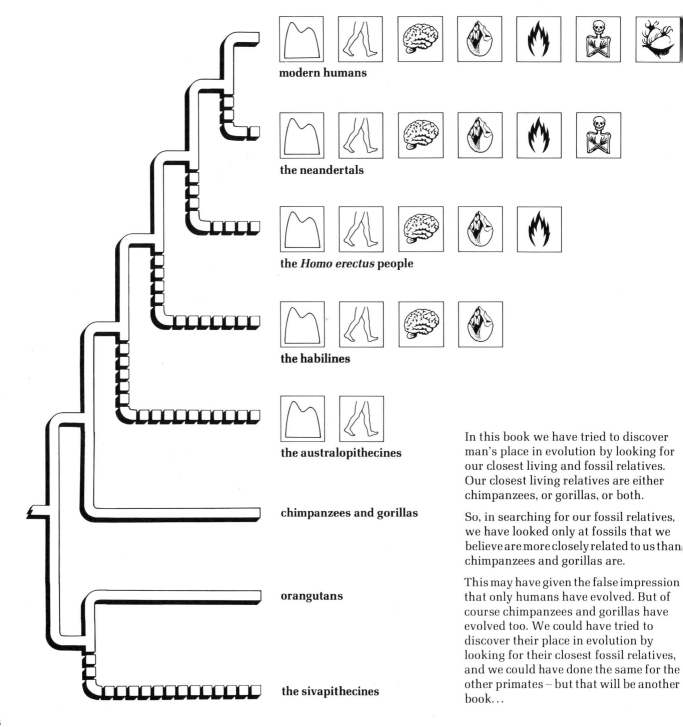

modern humans

the neandertals

the *Homo erectus* people

the habilines

the australopithecines

chimpanzees and gorillas

orangutans

the sivapithecines

In this book we have tried to discover man's place in evolution by looking for our closest living and fossil relatives. Our closest living relatives are either chimpanzees, or gorillas, or both.

So, in searching for our fossil relatives, we have looked only at fossils that we believe are more closely related to us than chimpanzees and gorillas are.

This may have given the false impression that only humans have evolved. But of course chimpanzees and gorillas have evolved too. We could have tried to discover their place in evolution by looking for their closest fossil relatives, and we could have done the same for the other primates – but that will be another book...

Further reading

. . . about man's living relatives

Monkeys and Apes by Prue Napier, Hamlyn All-Colour Paperbacks, 1971. A colourful little book that contains an amazing amount of information.

The Natural History of the Primates by J.R. & P.H. Napier, British Museum (Natural History) /Cambridge University Press, 1985. Lavishly illustrated introduction to our closest living relatives.

The Chimpanzees by Prue Napier, Bodley Head New Biologies, Bodley Head, 1974. For children.

In the Shadow of Man by Jane van Lawick-Goodall, Collins, 1971. A fascinating first-hand account of life amongst wild chimpanzees on the shore of Lake Tanganyika.

Gorillas by Colin P. Groves, Arthur Barker/Arco Publishing Company, 1970. An affectionate study of the life and behaviour of gorillas, both in captivity and in their natural surroundings.

. . . about evolution

Evolution by Colin Patterson, British Museum (Natural History), 2nd edition in preparation for 1992. An authoritative introduction to modern evolutionary theory for people who have little or no technical knowledge of biology.

Origin of Species, British Museum (Natural History) 1981. A beautifully illustrated introduction to evolution.

. . . about man's fossil relatives

Lucy: the beginnings of humankind by Donald Johanson and Maitland Edey, Granada, 1981. A highly readable, often provocative account of the events surrounding the discovery of 'Lucy' and her relatives.

Missing Links by John Reader, Penguin Books, 2nd edition, 1988. A very readable account of various 'fossil man' discoveries and the people who made them.

Bones of contention by Roger Lewin, Simon and Schuster, 1987. This book provides many revealing insights into the work of scientists studying human evolution.

The Cambridge Guide to Prehistoric Man by David Lambert, Cambridge University Press, 1986. A concise, well-illustrated survey of primate and human evolution.

Human Evolution: an illustrated guide by Peter Andrews and Chris Stringer, British Museum (Natural History), 1989. Using Maurice Wilson's excellent painted reconstructions, this book briefly surveys the last 35 million years of primate and human evolution.

. . . about the development of farming

A Natural History of Domesticated Mammals by Juliet Clutton-Brock, British Museum (Natural History)/ Cambridge University Press, 1987. This book looks at the evidence for the first domestication of mammals, and at the exploitation of wild animals now and in the past.

. . . for fun

Once Upon an Ice Age by Roy Lewis, Terra Nova Editions, 1979. A really good novel, combining fact and fiction. The reader can imagine the emotions experienced by 'hordes' of the past who discover fire and the tantalizing smells of cooked food.

The Inheritors by William Golding, Faber, 1973. A delightful fantasy about a group of neandertals meeting with another group of early human beings.

Dance of the Tiger by Bjorn Kurten, Pantheon Books, 1980. This novel, written by a scientist, tells the story of the first meeting of neandertals and modern humans in Europe.

Names of the fossils featured in this book

(The numbers in brackets are the catalogue or Museum numbers of these specimens.)

The early apes

p.28

From Rusinga Island, Kenya
Proconsul africanus

p.29

From Ad Dabtiyah, Saudi Arabia
Heliopithecus leakeyi

p.29

From Fort Ternan, Kenya
(FT 46,47)
Kenyapithecus wickeri

p.29

From St Gaudens, France
Dryopithecus fontani

Sivapithecines

p.31

From Yassorien, Turkey
Sivapithecus meteai

p.31

From Potwar Plateau, Pakistan
(GSP 4622/4857)
Sivapithecus sivalensis

p.31

From Çandir, Turkey
Sivapithecus alpani

p.31

From Chinji, Pakistan
(GSI.D.118/119)
Sivapithecus punjabicus

p.31

From Haritalyangar, India
(YPM 13799)
Sivapithecus punjabicus

p.31

From Hasnot, Pakistan
(YPM 13814)
Sivapithecus punjabicus

Australopithecines

p.34

From Taung, South Africa
Australopithecus africanus

p.35

From East Turkana, Kenya
(KNM-ER 406)
A robust australopithecine, referred to as *Australopithecus boisei*. Also called *Paranthropus boisei*.

p.35,46

From Olduvai Gorge, Tanzania
(OH5]
A robust australopithecine, originally named *Zinjanthropus boisei*. Now referred to as *Australopithecus boisei* or *Paranthropus boisei*.

p.35,46

From Swartkrans, South Africa
(SK48)
A robust australopithecine, originally named *Parenthropus crassidens*. Also referred to as *Australopithecus robustus*.

p.38

From Makapansgat, South Africa
(MLD2)
Originally called *Australopithecus prometheus*. Now referred to as *Australopithecus africanus*.

p.42,46

From Sterkfontein, South Africa
(Sts 5)
A gracile australopithecine, *Australopithecus africanus*, originally called *Plesianthropus transvaalensis*.

p.43

Hip bone from Sterkfontein, South Africa (Sts 14)
A gracile australopithecine, *Australopithecus africanus*, originally called *Plesianthropus transvaalensis*.

p.44,45

Thigh bone from Koobi Fora, Kenya
(KNM-ER 738)
Australopithecus species.

p.44

Footbones and big toe bone from Olduvai Gorge (OH8, OH10)
Originally called *Homo habilis*. Now sometimes thought of as an australopithecine.

p.45

'Lucy' from Hadar, Ethiopia
(AL 288-1)
Australopithecus afarensis

p.46

From Peninj, Tanzania
A robust australopithecine, *Australopithecus boisei*. Also called *Paranthropus boisei*.

p.46

From Swartkrans, South Africa
(SK 23)
A robust australopithecine, *Australopithecus robustus*. Also called *Paranthropus robustus crassidens*.

p.46

Jaw from Sterkfontein, South Africa
(Sts 52b)
A gracile australopithecine, *Australopithecus africanus*, originally called *Australopithecus prometheus*.

Habilines

p.50

From Olduvai Gorge, Tanzania
(OH7)
Homo habilis

p.51

From Koobi Fora, Kenya
(KNM-ER 1813)
Referred to as *Homo habilis*

p.51

From Olduvai Gorge, Tanzania
(OH24)
Referred to as *Homo habilis*.

Homo erectus people ——————

p.62

'Heidelberg Man' from Mauer, Germany
Homo erectus heidelbergensis. Also called *Homo sapiens heidelbergensis* or *Homo heidelbergensis*.

p.62

From Olduvai Gorge, Tanzania
(OH9)
Homo erectus, originally called *Homo leakeyi*. Sometimes known as 'Chellean Man.'

p.62

From Swartkrans, South Africa
(SK15)
Homo erectus, originally called *Telanthropus capensis*. Has also been called *Homo habilis*.

p.63

'Beijing Man' from Zhoukoudian, China
Homo erectus pekinensis, originally called *Sinanthropus pekinensis*.

p.63

'Solo Man' from Ngandong, Java
Homo erectus soloensis, originally called *Javanthropus soloensis*. Has also been named *Homo sapiens soloensis*.

p.63

'Trinil Man' from Trinil, Java
Homo erectus erectus, originally called *Pithecanthropus erectus*.

p.69

From Petralona, Greece
Originally thought to be a neandertal. Has been called *Homo sapiens heidelbergensis*, *Homo erectus heidelbergensis* or *Homo heidelbergensis*.

Neandertals ——————

p.72

From the Neander Valley, Germany
Homo sapiens neanderthalensis, originally called *Homo neanderthalensis*.

p.72

From Forbes Quarry, Gibraltar
Homo sapiens neanderthalensis, originally called *Homo calpicus*.

p.73

'Swanscombe Man' from Swanscombe, England
Usually called *Homo sapiens steinheimensis* or *Homo sapiens neanderthalensis*.

p.73

From Amud Cave, Israel
Homo sapiens neanderthalensis

p.74

From La Chapelle-aux-Saints, France
Homo sapiens neanderthalensis, originally called *Homo chapellensis*.

p.75

From La Ferrassie, France
Homo sapiens neanderthalensis

Index

A

Ad Dabtiyah, Saudi Arabia 29, 98
alpha-haemoglobin chain 19, 25
Amud, Israel 73, 100
apes 14, 29, 30, 32-33, 35, 41
    characteristics of 11, 18, 22, 28
australopithecines 34-47, 51, 53, 96, 98-99
    scientific names of 98, 99
    Australopithecus afarensis 45
    gracile australopithecines 46-47
    robust australopithecines 46-47

B

backbone 10, 16, 42
Balzi Rossi, Italy 85
'Beijing Man' 63, 66, 68, 74, 100
blood 19, 22
bones 16-17, 25, 42-45
'brain cast' 52, 64
brain size 37, 51-53, 64, 74-75, 94
Brassempouy, France 85
burial 74, 75, 77, 94

C

Çandir, Turkey 31, 98
carvings 85
cave paintings 85
ceremonies 72-77, 94
Chapelle-aux-Saints (La), France 74, 100
chimpanzees 17-19, 22-25, 28-29, 32-33,
    35-40, 47, 52, 53, 96
Chinji, Pakistan 31, 98
chromosomes 18-19, 24, 25
cladogram 14-15, 22, 24, 25, 28, 29, 33
closest relative 14, 16, 22, 23, 25
common ancestor 9, 14, 16, 18, 19
cultivated plants 86-87, 91
cusp patterns 28, 37, 38, 51, 64

D

diet 46, 79, 86
domesticated animals 86-87, 91
Dryopithecus (fontani) 29, 33, 98

E

ear bones, middle 10

East Turkana 34, 98

F

farming 86-91
Ferrassie (La), France 72, 75, 100
fire 62, 63, 64, 65-7, 74, 94
foot 44
footprints 44-45
Forbes Quarry, Gibraltar 72, 100
Fort Ternan, Kenya 29, 98
frontal sinus 17, 22, 32, 35
fur 10

G

gibbons 17, 18, 22, 28, 29, 33
gorillas 17-18, 22-25, 28-29, 33, 35-40,
    40-44, 47, 96
gracile australopithecines 46-47

H

habilines 50-59, 64, 94, 96
    scientific names of 99-100
Hadar, Ethiopia 45, 99
haemoglobin 19
hair 10, 25
Haritalyangar, India 31, 98
Hasnot, Pakistan 31, 98
'Heidelberg Man' 62, 100
Heliopithecus leakeyi 29, 33, 98
hip 43, 45
Homo erectus people 62-69, 74, 94, 96
    scientific names of 100
Homo habilis 59
    see also habilines
Homo sapiens 59, 69, 75
    see also modern man
Homo sapiens neanderthalensis 75
    see also neandertals
homologies 16-19, 22, 24
human beings 9, 28, 29, 32-33, 59, 96
    characteristics of 10-11, 17-19, 22-25,
    35-45, 47, 51-54, 65, 75-76, 84
hunting 56-58, 67, 79, 84-85, 89

I

incisors
    see teeth
ischial callosities 18

J

Jaws 28, 31, 38-39
Jericho 90-91

K

Kenyapithecus (wickeri) 29, 33, 98
knuckle walking 40-41
Koobi Fora, Kenya 44, 51, 99

L

Laetoli, Tanzania 45
language 84, 85
Lascaux, France 85
Lucy 45, 99

M

Makapansgat, South Africa 34, 38, 99
mammals 10, 18
man
    see human beings
Mauer, Germany 62, 100
middle ear bones 10
milk-producing glands 10, 18
model-making 80-81
modern man 59, 64-65, 68, 74-76, 84-91,
    94, 96
    see also human beings
molars
    see teeth
molecules 16, 19, 24, 25
monkeys 18, 40
muscles 42, 43

N

nails, toe and finger 10
names, scientific 59, 69, 98-100
Neander Valley, Germany 72, 100
neandertals 72-81, 96, 100
Ngandong, Java 63, 100

O

Oldowan industry 55
Olduvai Gorge, Tanzania 34, 42, 44, 46, 50, 51, 54, 55, 62, 98, 99, 100
'opposable' thumb 10
orang-utans 17, 22, 28, 29, 32, 33, 35

P

Peninj, Tanzania 46, 99
Petralona, Greece 69, 100
Potwar, Pakistan 31, 98
premolars
    see teeth
primates 10, 17, 19, 24, 25, 28
    characteristics of 10
problem solving 54, 59, 65, 89
Proconsul 28, 29, 33, 98

R

reconstruction of neandertal woman 80-81
relationships 14-19
Rift Valley, Africa 34
robust australopithecines 46-47
Rusinga Island, Kenya 28, 98

S

St Gaudens, France 29, 98
shoulder blades 11, 28
sitting pads 18
sivapithecines 30-33, 96
    Sivapithecus alpani 31
    Sivapithecus meteai 31
    Sivapithecus punjabicus 31
    Sivapithecus sivalensis 31, 32
    scientific names of 98
skull 17, 32, 35, 42, 47, 51-53, 64, 68-69, 74-75, 78
'Solo Man' 63, 100
Sterkfontein, South Africa 34, 42, 43, 46, 99
'Swanscombe Man' 73, 78-79, 100
Swartkrans, South Africa 34, 46, 62, 99, 100
symbols 84-85

T

Tabūn, Israel 73, 74, 80

Taung, South Africa 34, 98
teeth 16-17, 24, 25, 30, 38, 46-47, 64
    canines 25, 46, 47
    cusp patterns 28, 37, 38, 51, 64
    incisors (biting teeth) 10, 32, 35, 46, 47
    molars (chewing teeth) 11, 28, 29, 46, 47
    premolars 29, 37, 38, 39, 47, 51, 64
    wisdom 64
Teshik-Tash, USSR 73, 77
thigh bone 44, 45, 64
toolmaking 50-59, 64, 65, 74, 76
tools 50, 51, 54-55, 64, 74, 78, 91
'Trinil Man' 63, 64, 100

U

unique homologies 16, 22, 24

V

vertebrates 10, 16

W

walking 39-45
    knuckle walking 40-41
    upright, on two legs 37, 39-45, 51, 64

Y

Yassorien, Turkey 31, 98

Z

Zhoukoudian, China 63, 66, 100

Acknowledgements

Photographs
31, 38: Çandir jaw, Mineral Research and Exploration Institute of Turkey
44/45: footprints, John Reader, courtesy Dr Mary Leakey and the National Geographical Society.
45: Lucy, Cleveland Museum of Natural History.
48/49, 53: car factory, Fiat Auto (UK) Ltd
65: gas fire, British Gas Corporation
66: Choukoutien site, Institute of Vertebrate Palaeontology and Palaeo-anthropology, Academia Sinica, Peking
69: Petralona skull, Dr C. B. Stringer
70/71, 76: funeral of Sir Winston Churchill, Patrick Thurston, Daily Telegraph Colour Library.
76: wedding, D. W. Morbey.
85: Lascaux cave painting, Jean Vertut, courtesy Editions Mazenod (from *Prehistory of Western Art* by Leroi Gourham).
88: feeding sheep, *Farmers Weekly;* pigs, Spectrum Colour Library.
90: Jericho, British School of Archaeology in Jerusalem.
92/93: moon landing composition, NASA photographs AS11–446553, 16–107–17435 and 16–107–17438.

For their cooperation and help in the production of the following photographs, we should like to thank:

19: chromosomes, Dr Joy Delhanty, The Galton Laboratory, Department of Genetics and Biometry, University of London.
20/21: gorilla, Twycross Zoo.
60/61, 65: glass-blowing, Whitefriars Glass.
76: christening, Rev R. W. H. Nind, St Matthews, Brixton, London.
82/83/84//85: London Philharmonic Orchestra.
88: vegetables, Covent Garden Market Authority, London.